OCEANING

ELEMENTS   *A series edited
by Stacy Alaimo and Nicole Starosielski*

OCEANING

GOVERNING MARINE LIFE WITH DRONES

ADAM FISH

DUKE UNIVERSITY PRESS   Durham and London   2024

© 2024 Duke University Press
All rights reserved
Printed in the United States of America on acid-free paper ∞
Project Editor: Michael Trudeau
Typeset in Chaparral Pro and Knockout by BW&A Books, Inc.

Library of Congress Cataloging-in-Publication Data
Names: Fish, Adam, [date] author.
Title: Oceaning : governing marine life with drones / Adam Fish.
Other titles: Elements (Duke University Press)
Description: Durham : Duke University Press, 2024. |
Series: Elements | Includes bibliographical references and index.
Identifiers: LCCN 2023018773 (print)
LCCN 2023018774 (ebook)
ISBN 9781478030010 (paperback)
ISBN 9781478025801 (hardcover)
ISBN 9781478059011 (ebook)
Subjects: LCSH: Marine biology—Research—Technological innova-
tions. | Oceanography—Research—Technological innovations. |
Marine sciences—Research—Technological innovations. | Drone
aircraft in remote sensing. | Marine biology—Remote sensing. |
Marine sciences—Remote sensing. | Oceanography—Remote sens-
ing. | Information storage and retrieval systems—Marine biology. |
BISAC: SCIENCE / Environmental Science (see also Chemistry / Envi-
ronmental) | NATURE / Ecosystems & Habitats / Oceans & Seas.
Classification: LCC GC10.4.R4 F574 2024 (print) | LCC GC10.4.R4
(ebook) | DDC 577.7028/4—DC23/ENG/20231018
LC record available at https://lccn.loc.gov/2023018773
LC ebook record available at https://lccn.loc.gov/2023018774

Cover art: Photograph © Christian Miller / Ocean Alliance

This book is dedicated to my daughter, Io, the most reluctant nipper.

CONTENTS

During the fieldwork for this book (2015–22), I became a drone pilot, logging over a hundred hours of flight time in locations around the world: the United Kingdom, Iceland, Denmark, United States, Indonesia, Sri Lanka, and Australia. You could say I was seduced by the techno-utopian discourse of entrepreneurs like Chris Anderson, who left the editorship at *Wired* magazine to start the drone company 3D Robotics and evangelized: "We are as yet tourists in the air, briefly visiting it at great cost. By breaking the link between man and machine, we can occupy the skies" (C. Anderson 2017). Anderson's aspiration resonated with icons in my development: Stewart Brand's *Whole Earth Catalog*, Jacques Cousteau's marine conservation, Carl Sagan's vision of our interplanetary future—my thanks go to my mother, Jennifer Fish, for exposing me to these visionaries at a young age. Such early influences coalesce in this book on human-scale technology, the ocean, and becoming a mature planet.

My first drone experiment from 2015 to 2017 was designed to enhance "infrastructural literacy" (Parks 2009) by creating a visual travelogue of the undersea fiber-optic system that connects Iceland, the United Kingdom, and Denmark. The discoveries from this research foreshadowed findings in *Oceaning*—namely, that drones are entangled by what they engage and conditioned by the environments through which they fly. We lost control of a drone because of its reaction to magnetic fields emitted from black sand beaches in Iceland (or so we imagined), and I witnessed the communicative parallels between the networked technologies of drones and the sonar of pilot whales in the Faroe Islands. These flights and crashes showed the elemental contingencies of exploration (Fish and Garrett 2019; Fish, Garrett, and Case 2017a, 2017b; Garrett and Fish 2016). My appreciation goes out to Dr. Bradley L. Garrett for partnering in these and other projects and for introducing me to drone piloting. Páll H. Vesturbú was instrumental in visiting the internet's points of presence in the Faroe Islands. A visiting professorship at the University of Iceland, courtesy of Hafsteinsson Sigurjón, grounded me during the fieldwork in 2015.

A Leverhulme Trust Research Fellowship (2017–18), in the UK, enabled me to conduct ethnographic work into drone culture in northern Europe, Indonesia, and Sri Lanka. At the time, I was faculty at Lancaster University in the sociology department, a conducive place to study the politics of science and technology. My sincere thanks go out to Dr. Monika Buscher, Dr. Tim Dant, Dr. Bulent Diken, Dr. Bruce Bennett, Dr. Graeme Gilloch, Dr. Adrian Mackenzie, Dr. Corinne May-Chahal, Dr. Lucy Suchman, Dr. Imogen Tyler, Dr. Claire Waterton, Dr. Benjamin Neimark, Dr. Luca Follis, Dr. Karolina Follis, Dr. Brian Wynne, and Dr. John Urry. Releasing a video-enabled helium balloon with Dr. Bronislaw Szerszynski and somehow finding it in a backyard a hundred kilometers away was typical of the type of experimentation encouraged at Lancaster University. The Centre for Mobilities Research, the Institute for Social Futures, Security Lancaster, and the Faculty of Arts and Social Sciences at Lancaster University, as well as COST: European Cooperation in Science and Technology, deserve thanks for being early supporters of this scholarship.

During 2018–19, I navigated with drone inventors, humanitarians, and environmental activists as they piloted drones during crises. Environmental and land-rights advocates in West Papua, Indonesia, steered drones to record riverbank erosion caused by palm oil plantations. They documented their traditional arboreal gardens and surrounding peat swamps—all preemptive efforts to map their lands before the plantation owners rolled into the area, demanding land concessions. I assisted these operations, bringing pilots and engineers together with local activists and drones (Fish and Richardson 2022; Fish 2023). As a documentary filmmaker, I saw the immense possibility in this flying camera, while as an activist, I recognized how the drone could be a tool for radical transparency, a way to peer down into what geographer Doreen Massey (2004) calls the geographies of responsibilities and power. As an anthropologist, I was witnessing the emergence of a multisited public of cultural production, science, and technology. I would like to thank Dr. Irendra Radjawali and Hagorly Hutasuhut, both of the Bandung Institute of Technology, for their efforts in enlightening me about the potentials of atmospheric technologies in Indonesia.

While we were in Indonesia, the Agung volcano on the island of Bali erupted, and Aeroterrascan, an Indonesian drone company I was working with, donated their time, personnel, and technology. We operated drones over the volcano and looked into the crater to ascertain the severity of the eruption. What we saw encouraged the government to relocate seventy thousand people away from the epicenter. Later, I elevated drones over coral reefs in Bunaken National Marine Park in Sulawesi, Indonesia, documenting coral bleaching. Throughout these missions folks at Aeroterrascan—Dian Rusdiana Hakim and Feri Ametia Pratama—were generous with their time and my inadequate Bahasa Indonesia.

In Sri Lanka, I worked alongside conservationists defensively arming local farmers with drones so that they might frighten marauding Asian elephants (*Elephas maximus*) away from their fields and homes, saving themselves, their livelihood, and the elephants from a more lethal deterrent. The team at the Sri Lanka Wildlife Conservation Society, in particular Chandima Fernando, provided key insights into the potentials for drones in conservation.

The ideas in this book came together in the urban centers of Germany during a senior research fellowship at the Weizenbaum Institute for the Networked Society, at the Technische Universität Berlin, with Dr. Stefan Ullrich from 2018 to 2019 and a research fellowship at the Zemki Centre for Media, Communication and Information Research, University of Bremen, with Dr. Andreas Hepp in 2018. During this time I was fortunate to receive invitations from Dr. Johan Lindquist to speak at the University of Stockholm, Department of Social Anthropology; Dr. Tora Holmberg at Uppsala University, Department of Sociology; Dr. Patrick Vonderau at the University of Halle-Wittenberg; Dr. Paula Bialski and Dr. Götz Bachmann, then at Luephana University; and Dr. Philippa Lovatt at St. Andrews University. Helping to refine this book's concepts were additional talks at the News and Media Research Centre, University of Canberra; the Mobile Life Center at the Stockholm University; Technologies in Practice Research Group at the IT University of Copenhagen; the Social Anthropology Department at the University of Edinburgh; and the Institute for Advanced Studies in the Humanities and Social Sciences, National Taiwan University. I would like to express my gratitude to Dr. Beryl Pong for inviting me to give a keynote at the National University of Singapore's New Technologies Research Academy.

In 2019, I accepted a research position at the University of New South Wales, a university near the surprisingly undomesticated urban shoreline of headlands, bays, and beaches of Sydney, New South Wales, Australia. This book's conceptual framework began to concretize in Australia, where I found a research ecology steeped in environmental humanities and an ethnographic approach to multispecies studies (Van Dooren 2014; Rose 2011; Plumwood 2001). Sharks regularly patrol the shores while whales migrate north and south, and the coral of the Great Barrier Reef is two days' drive north. These animals are not distant specimens but strangely perceptible, in reach and in sight—perhaps too close for some. In Australia I have been fortunate to hang with drone-curious academics Dr. Edgar Gomez Cruz and Dr. Michael Richardson, who provided beachside opportunities for critical reflection on the purpose and potentials of these new technologies. The University of New South Wales's Scientia Fellowship program and Dr. Michael Balfour, in particular, generously provided me with the time to conduct this fieldwork, write this book, and participate in artist residencies. As I explored the drone's videographical capabilities, it became as much a writing as an artistic project. I was fortunate to work with two clans of brilliant artists at the Four-plus one: the Elements, Kinono Artist residency in Tinos, Greece, in May 2022 and the Field_Notes-The Heavens, Bioart Society, Kilpisjärvi Biological Station in Lapland/Finland, in September 2019. My thanks go out to the Open Eye Gallery in Liverpool, UK, for screening my film *Organic Machine* (2020) in January 2022 as part of the Look Climate Lab. I acknowledge those who screened my documentary *Crash Theory* (2019) about tumbling drones and collapsing ecologies. Likewise, I celebrate those who brought our *Points of Presence* (2017) documentary on information infrastructure in the North Atlantic, produced with Bradley L. Garrett and Oliver Case and music by Jon Christopher Nelson.

In Australia, I launched into a public of ocean drone hobbyists, scientists, entrepreneurs, and activists. I began conversing with people like Dr. Vanessa Pirotta, a marine biologist at Macquarie University who collects whale breath with drones; Jonathan Clark (of the Sea Shepherd Conservation Society in Australia) and Andre Borell (shark activist and filmmaker), who both advocate for the use of drones to alert swimmers of sharks' presence instead of using nets and hooks to kill and capture sharks; Gary Stokes, who flies drones and identifies illegal shark fishing in Timor-Leste and ghost nets near Hong Kong; Nora Cohen, previously of the Obama White House and now of Saildrone in San Francisco; and Dr. Karen Joyce, a coral drone scientist who started the crowdsourcing drone platform Geonadir. My thanks go out to these scientists and activists for their efforts in conservation and in enlightening me about their work.

Through interviews, I began to investigate the Sea Shepherd Conservation Society's long history of atmospheric and oceanic technologies in acts of direct action on the open seas. Their numerous television programs and feature-length documentaries reveal the innovative and sometimes terrifying applications of conservation technologies in the Southern Ocean to defend Antarctic minke whales (*Balaenoptera bonaerensis*) and the vaquita porpoise (*Phocoena sinus*) in the Sea of Cortez, Mexico. Dr. Pirotta's use of drones led me to Ocean Alliance, founded by Dr. Roger Payne, famous for his studies of humpback whale songs. I discovered through several conversations that, like Sea Shepherd, Ocean Alliance had long used a wealth of different technologies to get closer to and know more about whales. Shifting from shooting a crossbow at a whale to collecting skin and blubber samples, Ocean Alliance and their chief scientist, Dr. Iain Kerr, navigate drones to fly through the stinky exhalation of whales. To expand my understanding of how different drone technologies are affected by the elements, I visited the National Oceanic and Atmospheric Administration in Seattle and spoke with oceanographers Dr. Calvin Mordy and Dr. Carey Kuhn, who direct semiautonomous sea surface drones toward northern fur seals. I thank them and the many other drone pilots with whom I talked for their help in this project.

The ideas in this book have diffused across various other projects. Chapters 2 and 5 are particularly indebted to my 2022 article "Saildrones and Snotbots in the Blue Anthropocene: Sensing Technologies, Multispecies Intimacies, and Scientific Storying" in *Environment and Planning D: Society and Space* 40, no. 5 (October): 862–80; chapter 3 to my 2022 article "Blue Governmentality: Elemental Activism with Conservation Technologies on Plundered Seas" in *Political Geography* 93, art. no. 102528 (March): n.p.; and chapter 6 to my 2021 article "Crash Theory: Entrapments of Conservation Drones and Endangered Megafauna" in *Science, Technology, and Human Values* 46, no. 2 (March): 425–51. Gratitude to the editors and peer reviewers for the help on those articles.

Much appreciation for the conceptual assists given by Sarah Jane O'Brien, Courtney Berger, Nicole Starosielski, Stacy Alaimo, and Michael Trudeau. A hearty thanks goes out to my wife, Robin Fish, for enduring these escapades, for caring for our home and daughter, Io, when I was away, and for our partnership across these continents.

BEGINNING
INTIMACIES OF CONSERVATION
TECHNOLOGY

CATCHING A WHALE

Catching a humpback whale (*Megaptera novaeangliae*) is difficult. See-ing the spurt of its exhale, illuminated by the sun over a seaside cliff on a windless morning, is the easy part. From the eastern suburbs of Sydney, New South Wales, Australia, my eyes—first naked, then with binoculars—spot a humpback whale near south Bondi Beach, about two kilometers away, heading north. Monitor in hand, fingers on levers, heavy first-person viewing (FPV) goggles fitted around my head, I launch the drone on a long flight toward the splashing whale.

For five or so minutes, four razorlike propellers imperceptibly spin, temporarily lift, and carry an assortment of hardware, software, sen-sors, lenses, storage systems, and valuable footage. For this period, all that matters is staying aloft, not injuring another being, and executing the mission—to fly near a whale and film its behavior. Piloting the drone with FPV goggles is disorienting. On this sandstone cliff above a high tide, I get dizzy. I sit, then lie down, trying to calm my breathing. The water world inside the goggles is vibrant, detailed, and strangely peaceful. I scan the screen for little white bursts of a whale breathing or, better yet, breaching. Schools of Australian salmon (*Arripis trutta*), bottlenose dolphins (*Tursiops aduncus*), and jumping yellowtail kingfish (*Seriola lalandi*) swirl by as I travel to the whale; a dolphin mother turns to her side to take a closer look at the drone, redirecting her calves away from the buzzing, flying camera.

If I get there in time and the whale, weather, and technology coordinate, I might be able to capture video of a living whale. This spectacular video could document an individual, as whales are tracked by marine biologists according to the distinctive patterns on their flukes. Perhaps this whale has been photographed before by onlookers on a tour boat, another drone operator, or a whale scientist, and its images have made their way into an archive of whale pictures. Aided by artificial intelligence, this footage could be compared to a geographical database of whale flukes and might contribute to understanding this whale's migration. This is what I dream as I race at forty kilometers an hour over the sea, scanning for cracks in the blue horizon.

Suddenly, one kilometer away, the drone's connection to the remote control and goggles evaporates. The screen freezes, shifts to black and white, and goes blank. The drone is disconnected, hovering in place, or so I pray. This has happened countless times before, for reasons I'll never understand, and all I can do is hope it will find its way home. Otherwise, it has fallen and drowned, and with it the footage of the dolphins and fish, and no whale today. I resist the desire to take off the bulky headset; there is no point. The drone is too far away to see with my unaided eyes. So I sit in dark sensory deprivation, the once kaleidoscopic colors inside the goggles now an ebony void, listening.

Tension gives way to relief when I hear the propellers buzzing high above. Following its safety protocol, the drone automatically returns to where it departed, beside me on a headland of chalky sandstone. It lands with an alien precision. Emerging from the cavern of augmented reality, I take a deep breath and turn around to see that the whale—or at least its exhale—is still in sight. Joyful gasps and pointing fingers of a nearby couple reassure me that it remains in the neighborhood. With the exchange of a fresh battery, the goggled monitor returns to its multihued pixels, and I try to ignore the earlier threat of the drone sinking.

Zipping out to sea again, I follow the whale's pectoral fins as they beckon with playful waves. I make it. One hundred meters above the

FIGURE 1.1. Humpback whale breaching as seen by drone, 2021. Photo by author.

whale, I begin to record video, coordinating my piloting with its swimming. It dives. Its massive blue-black body fades to green and, as it turns underwater, it exposes its white underbelly and callosities before disappearing. Then, building force from the depths, it begins its breach (figure 1.1). Puncturing the surface with its full body, it appears to hover in the air before collapsing with a burst of white water. It is too far away to hear, but I imagine the sound of the thirty-ton animal smashing the sea. What I actually hear is the crashing of waves against the cliff below. Drone piloting is an experience not of disembodiment but of double embodiment—a seeing there and a being and hearing here. It is disorienting and profound.

My onshore body is barely mobile and slightly nauseated. Adrenaline soars, fingers twitch, and eyes patrol the digital seas. The drone stops recording; the memory card is full, complete with ten minutes of stunning footage of a whale slapping, diving, and porpoising or peeking out from the watery depths. It could be communicating, playing, or scratching callosities on the tension of the water surface—no one is quite sure why whales slap and dive as they do.

I am no longer able to record, so I simply travel with the whale for as long as my battery will carry me, coordinating my direction and speed with the whale's, a leisurely two kilometers per hour. Breathing, with no people around or evidence to collect, we just hang out. The drone and I work together to predict the whale's next move—a choreography of pilot, technology, atmosphere, whale, and ocean.

One likes to imagine that this is a happy whale. Its body is fat from feeding on krill in the Southern Ocean, and it's on its way to mating in warmer seas to the north. Humpbacks in Australia are protected—though if this is an older whale, it might remember being pursued by diesel-powered whaling vessels and explosive-tipped harpoons. The humpbacks of Australia have rebounded since the whaling moratorium of 1982. But other whales have not. The orcas (*Orcinus orca*) of the Puget Sound in the Pacific Northwest of the United States and Canada number merely seventy-four; two calves born in 2021 populate the Salish Sea with their chatter; and only 350 North Atlantic right whales (*Eubalaena glacialis*) exist—entanglements in ghost nets and lobster traps, ship strikes, and pollution spell their doom.

Whales are not the only marine species that are struggling. As many as 100 million sharks are killed annually, most for the cartilage in their fins, a tasteless ingredient in an expensive soup (Worm et al. 2013). Northern fur seals (*Callorhinus ursinus*) of the Pacific Ocean, once decimated for their luxurious hides, have yet to rebound. Corals are dissolving in an acid-rich sea, bleaching in a warming ocean. They are crushed under ships, snagged and dragged by nets, and blasted by dynamite fishers. Stories of oceanic demise abound. One could lament. I do.

Philosopher Rosi Braidotti, contemplating the loss of life of both the sixth extinction and the COVID-19 global pandemic, implores us to confront and yet be mobilized by this mourning:

> Considering the scale of the planetary suffering we are experiencing . . . it would be unethical to offer only theoretical tools: this is rather a time for solidarity, collective mourning, and regeneration. We need to pause to meditate on the multiple losses of both human and non-human lives, as well as deploy intellectual tools for further understanding and criticism. But over and above all else, an affirmative relational ethics is needed, driven by environmental principles, which combine more inclusive ways of caring, across a transversal, multi-species spectrum that encompasses the entire planet and its majority of non-human inhabitants. (2020, 28)

We can bemoan erasure, the individuals and communities extracted from a web of relationships, as well as exemplify an ethics of critique, creativity, and possibility about our collective predicament. So while my feelings about extinction are fused within its chapters, *Oceaning* is not about my dread of the ocean dead. Rather, in Braidotti's affirmative and practical spirit, this book examines the marine conservationists and activists who are inventing and deploying innovative technologies and techniques to preserve the lives of marine species. Neither sorrowful nor particularly hopeful, they emphasize technological action infused with scientific curiosity and activist passion. For these marine scientists and ocean activists, good tools can make a difference. They want their technologies to be useful for managing and controlling death and decline. This governance of life, they feel, is a form of caring for others and the future.

By any measurement the drone is a potent technology for exercising this care. It stretches the human body, its vision, mobility, and intentionality through the air, over the ocean, and into the animal's world. In its ability to see in greater detail as it watches the health of individuals, to move faster to intervene in the dangers these animals face, to disturb illegal activities, and to produce graphic images of animals thriving and dying, the drone is a transformative tool for an affirmative multispecies ethics. The data that drones gather—compelling sights, biological samples, and videos of animal behavior and illegal fishing—also change the piloting scientists and activists like me, electrified on a headland on the coast of Sydney.

Drone-collected images of emaciated whales, coral bleaching, hunting sharks, shark fin and fish-bladder poaching, abandoned seabird colonies, nesting sea turtle tracks, and the feeding habits of fur seals are saved on computer hard drives, sent to prosecutors, developed into scientific articles, or distributed on social media. But while the drone has revolutionized scientific data collection and the documentation of illegal fishing, translations of that information into positive impacts for marine lives are few. But they are developing. The work continues while climate and existential chaos reigns. This book examines the work of those pushing the bounds of what this technology can do to conserve marine life.

For that moment on the headland, though, I was not haunted by thoughts of extinction. It was the beginning of the humpback migration along eastern Australia, and I was sharing a moment with a whale via a drone. It was ephemeral and sublime. I longed to fly the drone closer, to see the contours of its body more vividly, to watch as it played and fed. But I would never be able to get close enough. No technology can transcend the species barrier. Regulating other species' survival is also beyond our complete control. Sensing technologies are deployed; scientists, conservationists, and concerned citizens monitor migrating whales; and the ocean and atmosphere shift their thermal and chemical states. Understanding and managing synergies across these living human and nonhuman elements is necessary yet ultimately partial. It will not be perfect, perhaps barely sufficient, but those of us who care try. I will upload my whale video to a database where it might improve the identification of this individual whale. Sometimes this is all we can do: watch from afar, with disturbances balanced by our intent to help.

INTIMACY

The drone expresses an ancient longing to lengthen normative human senses outside of our bodies. With its enhanced movement, speed, vision, and force from afar, the drone allows many people to move closer or see in higher resolution the world beyond our terrestrially bound bodies. Adopted into oceanography, the drone enhances proximity to marine life with consequences for understanding relationships between technologies, humans, and animals. *Oceaning* investigates the conservation possibilities of drones as they fly close to whales in the Sea of Cortez to collect viruses from their breath, follow starving seals instrumented with video cameras in the Bering Sea, frighten nesting terns in California, record coral bleaching and tracks of mother green sea turtles (*Chelonia mydas*) looking for a shore to lay their eggs in the Great Barrier Reef, and track sharks and the poachers who chase them, porpoises, and whales throughout the world. The book offers a proposal of technological intimacy that invites us to balance the networked connectivity provided by the drone with multispecies flourishing. With this closeness comes the aspiration to manage marine populations carefully yet incompletely.

Technologies like drones provide humans sensual extension that makes enhanced intimacy with animals possible. Intimacy, for technoscience and feminist scholar Kath Weston, is mediated by technologies that "draw people into new forms of embodied intimacy with themselves and with others" (2017, 10). The drone's sensual and motile extension enhances seeing, proximity, and the felt experience of closeness. This vision and proximity are not distinct qualities but are mutually reinforcing. Drone vision is a kind of touch, photons inscribing sensors with textures, a haptic-visual etching somewhere between sight and feeling (Ballestero 2019, 14). As they generate interactions between moving sensors and living bodies, drones increase the felt presence of physical proximity, visual resolution, and influential actionability.

Ocean anthropologist Stefen Helmreich writes that instead of "remote sensing," the kind of "intimate sensing" enacted by drone conservation is not a "detachment from nature, but of a pleasurable, technological immersion in it" (2009b, 142). The visual, embodied, and affective relations of intimate sensing invite a shift from an instrumental to an intrinsic valuation of life (Plumwood 2001). Animal ethicist Lori Gruen (2015) offers "entangled empathy" or perceptive caring that crosses the conscious, emotional, and species divide and recognizes the responsibilities that come with those relationships. Thus, intimate, entangled, and empathic sensing attends to "another's needs, interests, desires, vulnerabilities, hopes, and sensitivities" (Gruen 2015, 3). This human and animal intimacy may influence interspecies commitments of conviviality and concern (Braverman 2014; Youatt 2008). Companionship may follow (Haraway 2003).

The experience of enhanced adjacency felt through high-definition moving images and/or sensor contiguity makes possible a more reflexive, engaged, and ideally effective marine science and activism (Alaimo 2013). Intimacy returns affect to science (and names this impetus for passion in activism), reminding scientists to embrace their complicity and the living relations that foreground their methodologies. Intimate attention to nonhumans (e.g., scientific instruments and other animals) correctly positions these actants as integral to scientific and political production (Latimer and Miele 2013, 9). The drone leverages sensing from afar into possibilities for intimate caring (Paterson 2006; Puig de la Bellacasa 2009). This care results in efforts to manage marine life with technology.

TECHNICITY

Intimate contact is realized by the drone's ability to move from human bodies toward marine life. Technologies such as the drone are prosthetics of the normative human body and extensions of its senses. For archaeologist André Leroi-Gourhan ([1964] 1993), this theory of technicity begins with the evolution of modern human feet: upright standing freed hands to grasp and spines to become erect. With these adaptations the prehistoric human could see further, grasp and throw objects, hunt and eat calorically richer game, and grow a larger brain capable of complex social and linguistic development. The drone is the latest manifestation of this evolution toward escalating speed, locomotion, force, and interactions with other organisms. In collapsing the felt experience of space and enhancing intimate copresence with nonhumans, spears and nets, telescopes and microscopes, are kin to the drone.

For philosopher Bernard Stiegler, *technicity* refers to "originary prosthetics"—the original extension of the normative body (1998, 98–100). Likewise, for media studies scholar Marshall McLuhan, "all media, from the phonetic alphabet to the computer, are extensions of man [sic] that cause deep and lasting changes in him and transform his environment" (1969, 54). There is no cultural production, writing, or reading without technologies of depiction. This lean toward technological determinism can be tempered by considering communications scholar Manuel Castells's remark "the dilemma of technological determinism is probably a false problem, since technology *is* society, and society cannot be understood or represented without its technological tools" (1996, 5). Thus, it is impossible to be a modern human body without technologies for reading, writing, listening, seeing, grasping, and moving. Drones are a later and unique manifestation of this basic and foundational originary technicity.

Recent scholarship pushes against this normative reading of technicity, arguing that in its universalism it is ableist, privileging a body that is capable, willing, and interested in extending itself. As media studies scholar Jonathan Sterne writes in his autoethnography of physical impairment and voice augmentation technology, "Every body (and everybody) is situated historically, ecologically, and politically" (Sterne 2022, 11). Similarly, science and technology scholar Max Liboiron (2021) in their investigation into the ocean, pollution, and colonialism supplies vital insights into how universalism fails, and how practices have different impacts depending on the relations in which they are embedded. This research invites us to articulate the distinct bodies, political orientations, and environmental conditions that manifest from general patterns of technicity. In following this direction toward the particular while keeping a link to the universal, *Oceaning* offers a relative technicity. By *relative*, I refer to the classic cultural relativism of anthropologists Franz Boas (1940), Melville Herskovits (1955), and Clifford Geertz (1973) and the social constructivism of science and technology scholars Bruno Latour and Steve Woolgar (1986), which recognize in similar ways that knowledge claims are relative to their communities of practice.

Like technicity, however, relativism is not absolute in its atomism. The drone is an assemblage constituted not just by its relative emergence from

originary technicity and how it extends the sensing abilities of a normative pilot, but also by its unprecedented nature: the unique ways in which
it is deployed, the legal significance given to its output, the political orientation of its users, the disruptions provided by the sea and ocean, and
other unforeseeable criteria. Thus, relative technicity refers to how particular technologies develop and carry with them universal origins. The
relative technicity of the drone comes from the drone's originary technicity and how its affordances invite new ways of thinking and doing. In
Oceaning, references to *originary technicity* indicate the normative aspects
of technicity eclipsing time and space, while mentions of *relative technicity* refer to how originary technicity manifests through the drone for particular bodies and objectives.

Relative technicity is grounded in the embodied practices of drone
pilots, stretching their senses and bodies across the air and sea in acts
of discovery, scientific data collection, activism, and the sheer joy of flying. Several media and communication scholars have written about the
uniquely embodied acts of drone piloting. Media studies scholar Julia
Hildebrand (2021), for instance, deploys an autoethnographic approach
to drone piloting, expressing the felt experience of space, mobility, relationality, and legalities. Media studies scholar Maximilian Jablonowski
(2020) takes a similar phenomenological approach, showing how drone
piloting goes beyond the visual and toward the embodied telepresence I
experienced on my flight over the humpback whale. Media studies scholars Elisa Serafinelli and Lauren O'Hagan (2022) reveal the emergence of a
visual language from the embodied experiments of drone hobbyists and
their contortions of drone affordances.

Like these academic drone pilots, conservation drone pilots also challenge the bounds of the drone's relative technicity, reformatting the
technological affordances toward their goals of political interventions in
challenging elemental contexts. As *Oceaning* reveals, working from the
universal drive to eclipse space and time, these scientists and activists
embody the relative in the drone's originary technicity. The affordance-
driven, practice-specific, and politically motivated acts of piloting a conservation drone are factors that modulate technicity into nonnormative
manifestations. In oceanography, relative technicity is concretized in
drones that gather information about species distribution, animal movement, climate change, threats to protected areas, conservation effectiveness, changes in land use, and ecological degradation. Drones watch
coasts for poaching and other illegal activity, spot whales before ship

collisions, track plastic garbage, aid search and rescue, sample oil slicks to find their origins, identify nesting birds, document wetland restoration, find whales entangled in nets, and instill a sense of wonder for the ocean. Conservation drones fly in every conceivable marine location—and a list of their most impressive applications is necessarily partial. Benefits are tempered by the challenges posed by ocean swells and tempestuous weather, regulations, gaps in legal systems, faulty technologies, error-prone pilots, and the disruptions that drones cause for animals.

ELEMENTALITY

The play of the elements—wind and sunlight, swell and current, heat and opacity—on the drone is unpredictable. In a recent magisterial survey using drones, thousands of marine mammals were counted across four beaches in New South Wales, Australia (Kelaher et al. 2020, 72). Bottlenose dolphins, Australian cownose ray (*Rhinoptera neglecta*), white shark (*Carcharodon carcharias*), and bull shark (*Carcharhinus leucas*) were identified by a machine-learning algorithm. The scientists complained that their task was made difficult by sun glint—the blinding refraction of visual light bouncing off the ocean (Giles et al. 2020). The shimmering sea, its wavy state, and the clarity or opacity of the water all distorted the data. As these Australian drone oceanographers confirmed, the ocean is more than its liquid state, just as the atmosphere is more than its gaseous mixture. These scientists labor outdoors. In heavy boots or sandals, in sunlight and saltwater, they work from lapping shores or sailing on watercraft, their vision extended into the skies via drones. Drones operate in, through, and despite the elements. Exploiting the conservation efficacy of the drone's relative technicity requires working with and sometimes against this atmospheric and oceanic elementality.

Elemental thinking has captured the imagination of scholars of science, technology, media, geography, and anthropology, but its emergence is ancient (Papadopoulos, Puig de la Bellacasa, and Myers 2022). The Sicilian Empedocles (494–c. 434 BC) provides the first written account in a Western canon of the *rhizomata* (ῥιζώματα), or roots of earth, water, air, and fire. In *Timaeus*, Plato followed and referred to each of these roots as an "element," *stoicheion* (στοιχεῖον). Indigenous and non-Western peoples have long understood the animacy and agency of earth, water, air, and fire, but the elements as the West knows them may have taken shape with the chemistry of the Irishman Robert Boyle (1627–91), who rediscovered phosphorus in the seventeenth century; Frenchman Antoine Lavoisier, who identified oxygen (1778); and Russian Dimitri Mendeleev (1834–1907), who developed what became the periodic table—a more complete graph of the material universe than that proposed by Empedocles.

Nevertheless, as basic as the pre-Socratic categories are, they, instead of the more exact periodic composition of the air (e.g., 21 percent oxygen, 78 percent nitrogen, 1 percent argon, etc.) or seawater (by mass the ocean is 86 percent oxygen, 11 percent hydrogen, and 3 percent of chloride, sodium, sulfate, magnesium, etc.), remain the substances that are commonly understood and felt by humans and others. Elemental philosopher David Macauley identified this when he wrote, "the elements of modern chemistry are not the elements of our more immediate, somatic, cultural, or imaginative experience" (2010, 5). When we discuss climate change, for instance, we may refer to the parts per million (ppm) of carbon dioxide in the atmosphere (423 ppm as of July 5, 2023), but we—humans and nonhumans alike—feel it as heat trapped in the air and sea. The elements of earth, water, air, and fire remain both experientially and analytically central to living experience, including the working lives of oceanographers, drone operators, and marine activists.

Drone conservation is elemental. Its practitioners conduct "atmospheric anthropology" (Choy and Zee 2015, 217). They observe, intermingle with, and take part in the "turbulences and volatilities of ubiquitous air" with practices that are animated by the atmosphere's interaction with the ocean (217). In drone oceanography, this atmospheric anthropology blends with a "seascape epistemology" (Ingersoll 2016). As it does for Kānaka Maoli or Indigenous Hawaiʻians, seascape epistemologies inform the drone conservationist's movements through the elements and understanding of how the ocean and winds are interconnected, vibrant, and transient. Seascape epistemology is existential; it embraces "re-creation and de-creation" (Ingersoll 2016, 5–6). Onshore or offshore, sea and atmospheric "waves of knowledge" inform ways of becoming with technologies, the elements, and marine life (Ingersoll 2016). Drone conservationists fly toward atmospheric anthropology and voyage for a seascape epistemology. In the process, they engage in an "elemental politics" (Jue and Ruiz 2021) or more specifically an "aeolian politics"—an existential politics that is contingent upon the wind's interaction with the ocean and technologies of perception (Howe and Boyer 2015).

In addition to the air, drone conservationists must contend with the sea. The ocean's liquidity seeps into, corrodes, and complicates ocean science and activism as well as theory building (Peters and Steinberg 2019). Through evaporation, the oceans move into the clouds, fall on the terrestrial Earth, drain through rivers, and return to the sea. Seawater can become solid (ice) and air liquid (mist). The ocean creates winds and carries volatile substances that have effects far inland. Depending upon the perceiver, many "oceans" exist, emerge in response to the atmosphere, and can be perceived at various resolutions through sensing technologies. The elementality of the ocean and the atmosphere is ontological; it involves fundamental substances and their materialities, consistencies, viscosities, and fluidities.

Elements disrupt—causing turbulence, crashes, sinking, asphyxiation, or worse. But the same elemental forces that can kill also provide the conditions for lift and movement. Indeed, the elements carry the chemical building blocks for life. In this way, "the materiality of the oceans is not the bug but the feature" (Vehlken, Vagt, and Kittler 2021, 8). Elementality is as much about limitations as capacities: the varieties of movement, seeing, communicating, and living that the elements afford and refute (McCormack 2018; Peters 2015; Parikka 2015; Starosielski 2015). The drone's relative technicity toward moving and seeing faster is conditioned by elementality. Drones slip into being with these shapeshifting elements. The impacts of fluidity, saturation, buoyancy, reactivation, and corrosiveness humble what can be achieved through sensing technologies, governance, and regulation (Braverman and Johnson 2020; Jue and Ruiz 2021).

BLUE GOVERNMENTALITY

Despite the benefits that drones provide, many scholars are not convinced that attempting to conserve wildlife with drones is beneficial. Flown in antipoaching operations, drones are involved in "vertical militarisation . . . [and] conservation territorialisation from above" (Lunstrum 2014, 824). Unlike most practitioners, such academic critics do not believe that drones, when paired with terrestrial technologies (sonar, radar, electro-optic and thermal vision, satellite images, motion and seismic sensors, etc.), can contribute to life's flourishing (Johnston 2019; Snitch 2015). Indicative of the critique of drones in conservation is the work of geographer Chris Sandbrook (2015), who views drones in conservation as violating humans' privacy, safety, well-being, and data security. In his analysis, people are not agents but unwilling recipients of vertical repression under the guise of animal preservation. This scholarship has much to say about the harms of conservation on people and less to contribute to other species' abilities to avoid extermination. *Oceaning* rectifies this anthropocentrism by empirically documenting the contingencies, possibilities, and consequences of drone power—the capacity to govern life with drones.

My thinking on governance is inspired by recent work on the "California Foucault" (Dean and Zamora 2021; Wade 2019). During his period in California—Santa Barbara, Claremont, and Death Valley—historian Michel Foucault enjoyed going on walks in the desert and forest, conversing with acolytes, taking drugs, and resonating with liberal youth culture. Changes in his theories of power occurred after Foucault's time in California in the mid-1970s. The human subjects of Foucault's *Discipline and Punish* (1977), as generated by domination, gave way to the human subjects of the much-revised second and third volumes of *The History of Sexuality* ([1984] 1990), who use sex to find and explore themselves. Foucault explored his personal limits with hedonistic nightlife, sexuality, sadomasochism, Zen, LSD, and Taoism in California (Miller 1993).

Foucault told his students that California liberalism "is a whole way of being and thinking. It is a type of relation between the governors and the governed much more than a technique of governors with regard to the governed" (Foucault 2008, 192–93). Enacted through relationships more than instruments, this governmentality was not imposed but emerged between arrangements of technologies of control and technologies of the self. Building from his time in California, his skepticism about the failures of the socialist left in the aftermath of the May 1968 general strikes and occupations of universities in France, and his thinking on population and territorial governmentality in 1976–77, Foucault presented this governmentality as a preferable alternative to subject formation under either monarchy or socialism.

He conceived of this concept of governmentality during a transformative moment in technological and cultural history, for also during the 1960s and '70s in California was the convergence of New Age self-inquiry with networking technologies that provided the blueprint for the type of internalized biopolitical knowledge and control that is possible with small sensing technologies today (Turner 2006). It would be "anti-Foucauldian" to attempt to explain Foucault's theories by finding some "truth" in this biography (Miller 1993). But by referring to the California Foucault, I ask us to consider how his theory of governmentality might play outdoors, on the ocean, and in the open air.

From this socio-technical milieu, different forms of control and self-management became possible, including a "left governmentality" (Dean and Zamora 2021, 37). So instead of the disciplines of sovereignty and authoritarianism, control could be internalized, developed from within and in response to the prospect of external monitoring. This is governance not through punishment, but as a structuring of the possible. This left governmentality opens and limits the field of action—it is a governance of and through freedom. As such, subjectivity could be fashioned as an entrepreneurial project within an environment of autonomy. For Foucault (1996, 1998) as for the subjects in this book, governmentalities may construct modalities of autonomy for wild life, human and otherwise, to exist, not through coercion, but through conditioning.

This governmentality aligns with that proposed by Foucault's friend and fellow philosopher Gilles Deleuze (1992), as a mobile, flexible, pervasive, and personalizable form of self and social control. Not the enclosure of disciplinarity, this control is ambient, ecological, networked, and computational. With an oceanic analogy, Deleuze writes, "The disciplinary man was a discontinuous producer of energy, but the man of control is undulatory, in orbit, in a continuous network. Everywhere *surfing* has already replaced older *sports*" (1992, 5–6). This control is not conscripted and stationary but flows with and from the subject. Governed not by the rules of games, the surfer moves with the oceanic force, both individually motivated and proscribed by hydrodynamic flows.

Foucault's writing inspired significant development in environmental thinking. For instance, geographer Stephanie Rutherford's (2007) "green governmentality" is the zoēpolitical control over the flourishing of nonhuman life. In *Oceaning*, I advance "blue governmentality" as an azure oceanic and cobalt atmospheric spin-off of Rutherford's viridescent version. Advancing a cautious if qualified governmentality is heterodox in academia. But along with the scientists and activists depicted in *Oceaning*, a dedicated minority of philosophers, feminist ecologists, sociologists, historians, and geographers increasingly support this approach (Braidotti 2019a, 2020; Bratton 2021; Crist 2015; Hamilton 2017; Kopnina 2017; Malm 2018; Nail 2021). They might agree that the possibilities are too great and the existential threats too momentous to stymie the use of sensing technologies in conservation. I am wary of the potential blowback for platforming technological governmentality and the surveillance, policing, and discrimination it might empower—for any species. Unredeemable to many academics and activists, the use of a compromised force is seen as inherently unjust and discriminatory (Lowery 2016; Ransby 2018). Negative biopolitics exacerbates surveillance capitalism, racist technological redlining, digital political tribalism, and predictive gender discrimination (Chun 2021; Srinivasan 2019; Noble 2018). It is essential for the emergence and defense of just human societies that negative biopolitics continues to be critiqued and publicly resisted. In this book I seek to show how nonhuman blue governmentality avoids these inhumane errors.

Important scholarship in conservation geography personifies this critical view of biopolitics (Adams 2017; Braverman 2020; Lunstrum 2014; Sandbrook 2015). In this scholarship, gazing, measuring, and monitoring practices have an inherent discriminatory bias, and management is manifestly problematic. Ocean and legal scholar Irus Braverman (2020) contends that oceanic sensing and management evinces a negative biopolitics with unforeseen harms. We may be witnessing in drone conservation "a broader move towards robotic management not only in the ocean, but also in planetary governance writ large" (Braverman 2020, 148). These critics of conservation technology fear that quantification regimes that gain traction in the ocean might swell onto land, eroding privacy and inundating people with draconian surveillance.

For the drone conservationists in this book, however, not all modes of sensing and management are equivalent. Differences in species, context, and technologies matter. As far as we know, marine animals do not have the same corporeal risks, privacy expectations, social harms, and susceptible psychologies as humans. Drones used for conservation contrast with those used by militaries. Survival may require—not be challenged by—conservation technologies.

It is true that ocean conservation carries a historical orientation that draws from military technologies, continues species subjugation, and is largely Western in origin (Lehman, Steinberg, and Johnson 2021, 1). The practices and ideologies of conservation emerged from the quantitative and cartographic imperatives of Western expansion. And yet ocean conservation cannot be reduced to the histories of environmental, territorial, or mammalian exploitation. Rooted in but now bifurcated from these precedents, some forms of ocean conservation can be accurately analyzed without recourse through capitalism, industrialism, colonialism, sexism, racism, and militarism. Books on how oceanography concurs with these topics are many, persuasive, and extensively researched (e.g., Oreskes 2021). *Oceaning* offers a complementary treatise.

Drawing from the situated use of small-scale conservation technologies, my findings are grounded in political and scientific practices. I build from close observations of technological and elemental praxis to illuminate what kind of biopower over flourishing is possible. Blue governmentality is in league with concepts such as "blue legalities" (Braverman and Johnson 2020, 20) and "marine biopower" (Mirzoeff 2009, 290), which strive to articulate how the ocean and the governance of biopolitics converge. Blue governmentality pairs the exercise of biopolitical governance through human means (laws, regulations, technological affordances, etc.) with the contingencies of aqueous fluidity, atmospheric gusts, ecosystem perturbation, and technological slippage (Fairbanks et al. 2019; Lehman 2016; McNeill 2020; O'Grady 2019; Povinelli 2016; Squire 2016). Blue governmentalities are not focused on a singular, imagined ocean, so no universal applications of normative science or prefigurative activism are offered. Rather, multiple and relative governmentalities emerge from interactions between several technologies, seas, animals, and laws within a reactive and fluid elementality.

In attempting to preserve marine existence, blue governmentality weaves humans and technology into it. Geographer Jamie Lorimer makes a zoēpolitical argument for "living well with wildlife," a practice that requires science as well as "fences, rifles, and cameras, alongside legal, economic, and political technologies" (2015, 191). In this calculation, wilderness needs people's protection, and autonomous oceans require technologies. This would be problematic if it were not pragmatic. Blue governmentality is neither final nor total; elements, technologies, animal agencies, and the uncertainties of certain humans' laws stifle its enactment (Steinberg and Peters 2015). Despite these contingencies, blue governmentality aspires to manage vitality and ensure its continuity.

The paradox of blue governmentality is that with this drive to care for living organisms comes an attempt to accentuate biopower within a milieu of faulty technologies, elemental disturbance, oceanic lawlessness, and animal instincts. Ocean scholars Irus Braverman and Elizabeth Johnson point out that what they call "blue legalities" make possible "a mode of governing with care" (2020, 20). This careful governance is elaborated upon by design theorist Benjamin Bratton, who names it "positive biogovernmentality," which "is concerned not only with how life emerges or is made free, but also with how it can be repaired, reproduced, sustained, and preserved" (2021, 120). Like Braidotti, Bratton simultaneously considers the loss of biodiversity and the mortalities of COVID-19 when he argues for the life-supporting capacities of positive biogovern-mentality, writing, "Positive biopolitics is the rationalism of the living, not the dead. It is a political and philosophical commitment to the real against reactionary constructivism and traditionalist vitalism. The notion that sentience is just too mysterious to grasp or that the natural order is too sacred to fiddle with, and that this actually suffices as the basis of an effective and ethical medical policy, is the daydream of a comfortable class who does not live with the daily agencies of sewage landscapes and exposed corpses" (2021, 40).

While this is a vivid visualization of the blight endured by impoverished human beings, it also depicts the elementality and existentiality of marine species swimming in waters polluted by sewage and among the corpses of their kin. The enactment of positive biogovernmentality bears importantly on potential responses to global pandemics, climate chaos, the extermination of biodiversity, and the demise of ocean life. The activists and scientists featured in this book believe that the present threat of marine extinction makes more invasive, technologically driven, legal, and managerial practices necessary. Mutual thriving may demand policing and surveillance of marine species and those that would harm them. For most activists and scientists in this book, biological nature is distinct from human culture. They work to separate nature from culture.

NATURE REALISM

In this labor, these conservationists personify an old idea about the relationship between nature and humans. Culture as divisible from nature is central to the West's self-conception after the scientific revolutions of the sixteenth and seventeenth centuries. In this explication, nature is a primordial environment that is distinguished from the human subject because of God's grace and humans' earthly achievements. Thoughtful, articulate, reflexive, and rational—the human is the exceptional figure that is born opposite the organic. Humans are the subject with agency, the definitive signal out of the indistinct chatter of nature. The consequences are clear—if nature does not have agency, it can be controlled by the only species that does. The seeing, naming, measuring, and managing of conservation, thus, are a continuation of this thesis that nature and culture are distinct.

While conservationists personify this nature and culture dualism with their work to classify and govern marine flourishing, they also embody the idea that humans are dependent upon nature, that nature is a human category that enhances its control, and that what was once nature has become transformed by cultural practices. They understand that a conceptualization of nature as alien is neither accurate nor universal. Modern humans evolved from and yet are reliant upon natural processes such as riverine, atmospheric, vegetative, soil, and oceanic cycles for the life-sustaining nutrients they make available for human gathering. Humans and other species are interwoven in surprising and complex ways that reach beyond the convenient binary of nature there and culture here (Descola 2012). To represent this togetherness, we have "natureculture"—humanity and nature as a living concomitant system (Barad 2003; Haraway 2003; Latour 2017). Proponents of natureculture are simultaneously humble and self-dignifying: humans are materially from and dependent upon nature while also capable of classifying and thereby inventing "nature," a process that emboldens its domination. The drone conservationists in *Oceaning* reject the stoic fatalism of the end of nature while understanding modern humans' role in both its demise and its preservation. These activists and scientists are proponents of what I call nature realism. From this perspective, nature is distinct and needs to be made more so through technological, geographical, and political means.

The subjects of *Oceaning* do not accept the erosion of nature as a thing in itself or into a contrivance. The scientists and activists in this book are driven by a belief that technology can support life, and that the technosphere can regulate some aspects of the biosphere. This does not make them ecomodernists who think global resilience can be geoengineered, bringing about a technocratic "great Anthropocene" (Hamilton 2016). They are far too aware of planetary boundaries, elemental constraints, existential challenges, political perturbations, human hubris and ignorance, and technological limitations to be merely "good managers of the standing reserve" (Crist 2013, 144). Rather, these scientists and activists are nature realists who focus on empiricism over speculation, ontology over epistemology, and pragmatism over cynicism—what is and what can be done.

An example of this nature realism can be found in the appropriately titled article "Underestimating the Challenges of Avoiding a Ghastly Future," which reports on the natural science of biodiversity loss and the social science of political impotence to avoid "the erosion of ecosystem services on which society depends" (Bradshaw et al. 2021, 1). Cowritten by scientists such as Paul Ehrlich and conservation theorists like Eileen Crist, this critical realist approach to natural resiliency—a praxis that is both scientific and activistic or what ecophilosopher Leigh Price (2019) names "deep naturalism"—builds on the complementary capacities of both nature and culture. Nature realism blends critical realism with deep naturalism into an existential and biocentric ideology for biodiversity conservation. The technicities and legalities of blue governmentality are ways of segregating nature from culture and generating ecological resilience that relies upon neither artificial nor natural selection alone.

Nature realists agree with natureculturism on key points: some aspects of nature are socially constructed; humans—some more than others—and their technologies, industrialism, and politics affect nature. The relative technicity that is applied in *Oceaning* through the drone seeks to strategically differentiate nature from culture in distinct marine and political contexts. Nature realists envision not the semiotic, technical, or political hybridization of nature and culture but rather the enfranchisement of its otherness through contingent technical and political means. Instead of natureculture, I call this ocean/culture.

OCEAN/CULTURE

According to natureculturists, nature and culture are the ultimate unmarked categories—the unquestioned stand-in for absolute oppositional realities. The contrast of nature representing on the one hand all that is not human, and culture on the other hand typifying all of humanity, is false, they contend. The two terms are one and the same, and either can refer to the other. Hence, Latour (2017) writes "Nature/Culture" with the forward slash representing an ambivalence toward the isolated existence of either nature or culture. The forward slash in ocean/culture in *Oceaning* is the opposite of this equivocation. In Boolean logic, the slash represents not a negation by the other but rather that the two are unique and interdependent entities. Where natureculturists reject it, nature realists strategically reify ocean/culture duality.

Nature realists take culture and nature as together but distinct, a state of "dependence *and* difference" (Malm 2018, 55). The forward slash of ocean/cultures symbolizes a divided togetherness, linking yet segmenting oceans and cultures. Similarly, sociologist Joanna Latimer (2013) does not see nature and culture, material and discourse, and noumena and phenomena as indistinguishably fused. Rather, she witnesses their woven paths running alongside each other, forming a helix of separate yet intertwined strands. When oil spills on the ocean surface, the ocean is not subsumed by the hydrocarbon industry. The two liquids remain distinct. The same is true with ocean/cultures. Far from fusing oceanic nature and some human's ambition into a mixture of mind and material, the care of nature realism exercised through the drone's relative technicity attempts the opposite. Through intimate contact, nature realism tries to separate industrialized humanity from marine existence.

Ocean/culture is inspired by environmental historian Andreas Malm and his thinking on "property dualism," the ontological rootedness and differences shared between nature and culture. They are of the same substance but are distinct in quality (Malm 2018, 59). This shared origin does not make nature and culture the same. He argues, "Nature is real; nature and society form a unity of opposites; society is constructed" (156). Indeed, this asymmetry makes distinguishing the two necessary. For it is the who (culture) that is transforming the what (nature). Malm's nature realism carries with it a call for political action. It asserts that now that humans are nearly everywhere, and much of nature has become more cultural, we must care for the nature that remains (Puig de la Bellacasa 2010). Nature realism proposes an ethic of justice for the ocean/cultures around us (Srinivasan 2013; Tschakert et al. 2020).

The drone oceanographers and activists detailed in this book believe that drones can contribute to the managed flourishing of the autonomy of whales, sharks, seals, turtles, and coral; stop illegal fishing; and slow extinction. Their goal is blue governmentality—the use of technologies to govern marine life. The drone is their tool for extending nature realism into the atmosphere and over the sea. This is an example of the practice of *oceaning*, the work of rendering oceans knowable through experience. Seafarers, fisherfolk, oceanographers, surfers, and the like are engaged in oceaning: physically navigating the textures, complexities, and dangers of the ocean. The term draws our attention to how oceans materialize cultures and how cultures socialize oceans. As many divers, sailors, marine scientists, and swimmers have confessed, with experience and awe often come respect and care. In this book, oceaning is political labor with drones that is inspired by nature realism and represents the leading edge of how humans differentiate oceans from culture via the contingent effort of blue governmentality. Never complete, this blue governmentality seeks to protect marine animals from poaching, overfishing, and extinction. *Oceaning* follows the practices of drone activists and conservationists as they fly and float closer to marine organisms in efforts to differentiate the vitality of the oceans from the effects of industrialized humanity. In the process, oceans and cultures come into close contact but ideally never meld into natureculture.

NATURECULTURISM

Ocean/culture is distinct from three versions of natureculturism: *nature constructivism*, wherein nature is semiotically fabricated; *natureculture materialism*, the argument that natural and anthropogenic functions have fused; and *political natureculturism*, the theory that articulations of nature are inherently iniquitous. Ocean/culture is not a conflation of nature and culture but a strategic differentiation of oceans and cultures, a process that can be built from technologically enhanced intimacy and can lead to new modes of human and more-than-human coexistence.

At just over 1.1 teratons, anthropogenic mass now surpasses biomass on the Earth. This includes, by weight, twice as much plastic as all wild terrestrial and marine animals combined (Elhacham et al. 2020). Natureculturists might argue that the scientists writing this report in *Nature* make a categorical error. The differences between anthropogenic mass and biomass is semantic—biomass *as* we know it depends on *how* we know it. "*There are no such independently existing objects with inherent characteristics,*" feminist physicist Karen Barad emphasizes, channeling philosopher Immanuel Kant. "In essence, there are no noumena, only phenomena" (Barad 2003, 816–17). According to this *nature constructivism*, nature does not consist of noumena—that is, objects in themselves; rather, nature consists only of phenomena or representations (816–17). Because things or materiality and speech or discourse are interwoven, categorization itself renders the "natural" amenable to human manipulation. How humans are a part of a planetary biomass and change and exploit it begins with how we observe and know it (Latour 1993). As media makers, the drone ocean scientists and activists in this book participate in this nature constructivism—inscribing meaning to fish, whales, and the sea. But they go farther, leveraging their animal stories into the means for survival (Van Dooren and Rose 2016). In this case, fish, whales, and the sea are noumena, not merely representations.

Natureculturism spans a spectrum from Barad's nature constructivism—that ways of knowing are ways of being—to *natureculture materi-*

alism, which argues that nature-as-noumena has ended. Natureculture materialists may point to zoos, wilderness parks, ecotours, genetic engineering, anthropogenic climate change, geological signatures of the Anthropocene, and so on as markers of transition from nature as a foreign other to nature as the cultivated self (Castree 2014; Kirksey 2015; Latour 2017; Purdy 2015). An extreme version of natureculture materialism is Braidotti's menagerie (2013), which is populated by "creatures of mixity or vectors" (73) such as "bioluminescent jellyfish and coral, 'Biosteel' goats . . . transgenic fruit flies [and] 'Atomic Age Rodents'" (2019a, 80). For natureculture materialists, nature and culture have merged into an "emergent ecology," a mixture of endemic and exotic species, Darwinism and industrialism, natural and artificial selection (Kirksey 2015). From this perspective, our most sensible and pragmatic approach is not mourning extinction but learning to live and die with what remains (Scranton 2015; Tsing 2015). The ocean scientists and activists in *Oceaning* oppose natureculture materialism, advocating instead for dividing nature and culture.

Finally, *political natureculturists* combine the semiotic containment of nature constructivism and the hybridization of natureculture materialism in a critique of how the concept of "nature" operationalizes ecological injustices. For these scholars, not only "nature" and "wilderness" but also "populations," and even "species," are socially constructed by Western elites for the liquidation of resources or as pseudoscientific curiosities (Cronon 1996; Werkheiser 2015). "Nature" is phenomena to be exploited as noumena. Political natureculturists are correct in identifying inequities in nature conservation—the fact that wilderness parks were only possible after the relocation or genocide of Indigenous folk, for example. Traditional owners, First Nations, and other Indigenous people should be able to reengage these landscapes, and their ecological knowledge should be integrated into management strategies. But these are ultimately problems for humans. The ocean scientists and activists in this book believe that the present threat of marine extinction makes more invasive, drone-driven, legal, and managerial practices necessary to protect nonhumans. Mutual thriving for both some humans and some nonhumans may require parks, policing, and surveillance. Instead of the semiotic enchantment of nature constructivism, the hybridity of natureculture materialism, or the anthropocentrism of political natureculturism, the ocean scientists and activists in *Oceaning* advocate for ocean/culture, a living marine world that is made materially and semiotically distinct from culture through the use of technologies, laws, and management.

This book combines the ecological and the materialistic with the political and applied, an approach that could be called ecomaterialist, or scholarship "that conjoins thinking the limits of the human with thinking elemental activity and environmental justice" (Cohen and Duckert 2015, 4; see also Oppermann 2019). As a work of personal technological experimentation, this is a type of "field philosophy" that is inspired by the interdisciplinarity of nineteenth-century natural philosophy and its experiments with then-recent technologies of perception to offer exploratory epistemologies about ecological relationships (Kieza 2020; Wulf 2015). Field philosophy is philosophy out of doors with prototypical technologies, practitioner communities, and applied ambitions (Ingold 2008). Embedded within the deeds and words of drone workers, this ecomaterialist field philosophy advances, refines, and challenges the ideology of its subjects whose nature realism combines environmental justice with the critical realism of applied science. Through the affordances of relative technicity and elementality, the practice of drone oceanography generates multispecies intimacies. The consequences for marine organisms are described by the control of blue governmentality and the fluid state of ocean/cultures. These concepts encode my attempts to synthesize academic philosophies on power and nature with the field labor, utterances, published scientific works, and hopes of oceanographers and marine activists.

Incorporating conceptualizations of technological capacities, the elementalities of atmospheres and oceans, animal agencies, and the contingencies of human, media, law, and enforcement—the chapters in *Oceaning* build on each other, showing the facets and phases of drone-driven conservation. For narrative effect, chapters 2–5, which did not involve participatory fieldwork, were written as if I were present for the events. Chapter 2, "Technicity," connects Leroi-Gourhan's ([1964] 1993) theories of the origins of technology to drone conservation labor and the increasing proximity between scientists and whales. The case study is Ocean Alliance, based in Gloucester, Massachusetts, which flies drones through the misty exhale of blue whales (*Balaenoptera musculus*) and gray whales (*Eschrichtius robustus*) in the Sea of Cortez, Mexico, to gather biotic data about whale health. Here, drone intimacy is forged through attention to whales and the drone's ability to touch their microbial exhaust. Integrating scholarship on touch by feminist Eva Hayward (2010) and scholarship on technologies of atmospheric attunement by anthropologist Kathleen Stewart (2011), this chapter develops an understanding of the delicate choreography of navigating a drone through a cloud of whale exhale while on a drifting boat in gusting wind, revealing the elementalities that make intimacy and care possible.

Enabling faster, more efficient, cheaper, safer, and more democratized movement across oceanic conditions than helicopters, drones allow conservation activists to pursue poachers. Chapter 3, "Elementality," follows the atmospheric and oceanic activism of the Sea Shepherd Conservation Society, a direct-action environmental organization that works to stop whale poaching in the Southern Ocean and porpoise killing in the Sea of Cortez. The major obstacles—but also the primary enablers—of their work are the elements themselves. Floating, flying, and sailing in pursuit of poachers is made possible by the atmosphere and the ocean's fluid states. This chapter gathers theoretical insights from marine geographers Kimberley Peters and Philip Steinberg (2019) to situate Sea Shepherd's use of technologies within the sea's liquidity. The activist drone in "Elementality" does not render the ocean's ontology, inhabitants, and politics into objectified abstractions. Rather, focus on conservation technologies shows the challenges of caring for marine species by controlling illegal fishing. In these acts of vigilante enforcement, blue governmentality assumes its limited capacities as compromised by the elements and the vagaries of prosecution.

Building on the previous chapter's investigation into poaching is chapter 4, "Governmentality," which stays with Sea Shepherd and explores the next stage in interdiction, the legal application of drone-derived data to not only disturb but also prosecute illegal fishing. This chapter examines a 2017 conservation mission to use drones to collect evidence of an illegal slaughter and transshipment of endangered scalloped hammerheads (*Sphyrna lewini*), sawfish (*Pristis* spp.), and tons of other shark-like fish, from the waters surrounding Timor-Leste and the Galápagos National Park, Ecuador, to China. Enforcing blue governmentality is immensely difficult. Diverse elemental, technological, and legal challenges are constant. Regardless, opponents of conservation governance argue that it is discriminatory against humans. Deploying a range of scholars from conservation theorist Helen Kopnina (2016) to the words of Sea Shepherd activists who were engaged in this operation, such as Teale Bondaroff, Gary Stokes, and Paul Watson, I advance the biocentrism of blue governmentality. The mortal consequences for nonhumans should blue governmentality be abandoned because of what it might mean for human populations are made clear through the efforts of Sea Shepherd to stop shark fin fishing.

With the foundation of blue governmentality—intimacy, technicity, and elementality—defined in the previous three chapters, the following chapters dive deeper into what conservation can be achieved with drones. First, the responsibilities of drone intimacies require meaningful analysis of data followed by communicating its significance. Chapter 5, "Storying," offers strategies for building scientific, artful, and multispecies narratives that dignify the intimacies that drones make possible. Both representing and performing this argument, this chapter tells the story of the National Oceanic and Atmospheric Administration's Pacific Marine Environmental Lab in Seattle, Washington, as they investigate the diets of northern fur seals by capturing them, collecting blood samples, attaching video cameras and dive trackers to their bodies, and following them and their prey, walleye pollock (*Gadus chalcogrammus*), with sea-surface sailing drones. A seal story with drones and melting ice as supporting characters might textualize survival in a warming Bering Sea. Such stories could advance life-sustaining conservation and enact a fitting reciprocity for the seals, whose time, blood, and energy are taken during their pursuit and capture. By connecting insights on narrative from field philosopher Thom van Dooren (2020; Van Dooren and Rose 2016), marine philosopher Serpil Oppermann (2019), and animal ethnographer Deborah Bird Rose (2011), I argue that storying constitutes reciprocal relationships between humans and nonhumans. The affirmative aspects of multispecies flourishing are unlikely without these nautical tales.

The species investigated throughout this book—whales, seals, sharks, porpoises, pollock, terns, sea turtles, and coral—are threatened by loss of habitat, pollution, overfishing, climate chaos, deadly accidents, or poaching. Drones are only effective at intervening in these troubles if they are in the air. But like the animals they investigate and as an experimental technology that is acted upon by elemental turbulence, the drone is often a subject falling from the sky. In chapter 6, "Crashing," close examinations of a crashing drone that forced a colony of elegant terns (*Thalasseus elegans*) to abandon their nests and the threat posed by crashing drones on an orca pod, both on the American West Coast, make evident ecological and technological fragility. Following Barad's (2007) understanding of how technologies and phenomena cocreate and science and technology scholar Steven Jackson's (2014) writing on technological repair, "Crashing" focuses on the recuperative work that is needed to strengthen the mutually beneficial relationships between ecological and technological resilience. Responding to the crashing drone and collapsing species with care and repair is necessary to reverse the demise of extinction. By examining drone failures, this chapter reveals one of the breaking points of blue governmentality and its aspiration of control.

Following the failed politics studied in "Storying" and the falling drones of "Crashing," chapter 7, "Living," offers a prototype of a functioning blue governmentality. Unprovoked shark bites are more frequent in Australia than anywhere else in the world. For over eighty years, the state governments of eastern Australia have dropped nets and baited hooks along many of its beaches. These interventions are designed to catch and kill large sharks, not stop them from approaching the shore—as many beachgoers assume. Shark advocates utilize their bodies, drones, and the media to present to audiences the brutality of these killing systems that also catch whales, turtles, stingrays, birds, dolphins, and, in one sad instance, a young boy. Following the work of shark geographer Leah Gibbs (2018, 2021), I argue that vulnerable yet strong human and shark bodies sharing space invite a way of being-with sharks that opens up the potential to replace the nets and hooks with drone surveillance. Toward this goal, the governments of New South Wales and Queensland have for several years financially supported an experimental program of flying drones to watch for sharks and alerting swimmers of their presence. The activists and those that support this use of drones advocate for a distinctive sense of space, not one which falsely separates shark and human territories, but rather invites a practice of coexistence with sharks. Instead of partially separating the shore between shark and human territories as the nets and baited hooks do, the drone shark surveillance program uses technology to nudge humans toward less fatal ways of sharing the sea, enabling sharks to swim more safely with people.

Like my drone on that Sydney headland hunting for humpback whales, many drones feature an automatic "return to home" function as a failsafe against a range of technological, elemental, and human errors. Insufficient battery or a loss of connection with the remote control will trigger this repatriation. With the simple push of a button, the pilot can summon the drone home if they perceive a danger, lose control, or simply want to abort the mission. After seven chapters of increasingly distant yet intimate flight, chapter 8 comes home, reconsidering how conservation technologies and their enhanced intimacy and capacity to invoke care and control alter the relationship between nature and culture. In "Coral/Cultures," I draw from my aerial and underwater drone survey of a protected island in the southern Great Barrier Reef in Queensland, Australia, to explore the limits of blue governmentality while documenting coral and green sea turtles. The fieldwork here, as uncomfortable human explorers in a marine park during seabird and sea turtle hatching season between tropical cyclones, shows the wild actualities of an ocean/culture, legally defended yet fiercely different from human culture. We were happy to return home.

But there is no returning to a time before the Blue Anthropocene. Like it or not, the technologies are part of conservationists' repertoire. They are tools for an incomplete preservation of the autonomy of marine life. Oceans are encultured by technology, yet technology is deployed to separate oceans from cultures—this is the paradox of ocean/cultures. The oceans are increasingly surveyed by drones and drone-like sensors, gliding the littoral air, skimming and bobbing on the surface, revolving in low Earth orbit, and floating through the vertical and horizontal columns of the seas. Conservation technologies more sophisticated than the drone are on the horizon. More desperate efforts at caring, protecting, and managing will arise, and most populations will continue to decline while a few may prosper. Although a more complete blue governmentality of the future may prolong certain marine life, it will also entrap organisms in technological dependencies (Giraud 2019; Hodder 2018). At the same time, the efficacy of controlling marine biopower will be tempered by unanticipated shifts in oceanic and atmospheric conditions, faulty technology, novel animal behavior, and the evolving evasiveness of illegal and unregulated fishing. The management of life stops when our care for each other ends. Conservation and its use of technologies provides a blueprint for an art of living together, while allowing others to be, as they are, distinct from us. *Oceaning* dives into the relative technicity of one conservation technology, the drone, considering how it is configured by the elements, and evaluates its possibilities and consequences for making the ocean that remains distinct from the culture to come.

2

TECHNICITY
TOUCHING WHALE EXHALE
WITH DRONES

ON THE WATER

It is 2016, and Ocean Alliance's team of five whale researchers drives through elephant cacti (*Pachycereus pringlei*), mudflats, salt ponds, and mounds of scallop shells before arriving at their research camp on the Sea of Cortez, near San Ignacio, Mexico. They gather their equipment in open skiffs and set out on a windless morning. After sailing for several hours, they turn off their outboard motor. Beginning with the late founder Roger Payne's early work recording humpback whale songs with hydrophones (distributed as the 1970 unexpected bestseller *Songs of the Humpback Whale*), Ocean Alliance has striven to find better ways to hear whales. Science manager Andy Rogan is onboard and listening for:

> the sound of the largest animal on the planet taking a breath: the "blow" or exhalation of a whale. It can be a difficult sound to explain. At its most practical, it alerts us to the presence of a whale: you often hear a whale blow long before you see it. Indeed,

with the mighty blue whales we have studied in the Gulf of California, you can hear the blow from well over a mile away! Sometimes, you also hear the whale inhale as well. This is more common with distinct species, such as fin whales, but is a wonderful sound—the sound of a vast cavern filling with air. It also means, of course, that there is the potential for a sample, which is the whole reason we are in Alaska or Mexico or Gabon in the first place. It is always great to hear a blow and then subsequently hear Iain [Kerr] exclaim "BINGO" or "oh, wow, the drone is covered in snot!" (Rogan n.d.)

As the oceanographers float, they listen and watch for exhale blasting from the sea surface. They can identify whale species by the shapes of these cloud exhales. Southern right whales' (*Eubalaena*) blow is V-shaped; humpbacks spit out shots three feet high; and gray whales (*Eschrichtius robustus*) expel moisture in the shape of a "love heart" (Giggs 2020, 78).

Upon the sound of the blow this morning, scientists, technologies, and marine species move into action. The team sails to within one hundred meters of a gray whale. Chief executive officer of Ocean Alliance and drone operator Iain Kerr pilots the drone to fifty feet above it and, with a black blanket draped over his head to minimize glare on the tablet that relays live video through the atmosphere, he lowers the drone to four meters once it is directly over the blowhole. He pays close attention to the direction of the wind and the angle of the whale's body. Kerr takes a brief breath, attuning himself to how many seconds lapse between the whale's exhales, and to the cetacean's speed and direction, to predict where the next exhale will appear. The drone's speed, mobility, and vision enable him to position it inside the whale's spray, drowning the petri dishes that cover its shell in snot. Elements, animals, and technologies conspiring—that is his hope.

He dips the drone at the right moment to two meters above the whale's blowhole. After a messy spray, its petri dishes are moist with blow. "Bingo!" Kerr indeed exclaims, and pilots the drone back to the skiff. A damp, slightly stinky veil of mucus coats the drone. The sample is capped and frozen. At some point in the future, it is expected, the whale snot will undergo polymerase chain reaction to discern its DNA constitution and the scientists will find out what lives inside this whale. This molecular intimacy is "the embodiment of data as mist" (Calvillo and Garnett 2019, 343) where elemental vapors constituting existential livelihoods

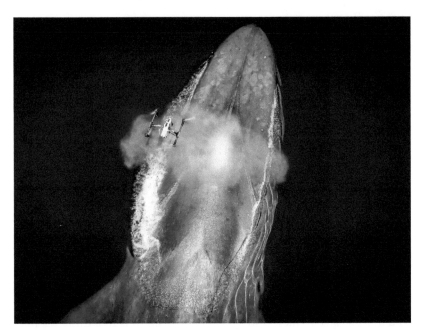

FIGURE 2.1. SnotBot collecting exhale from a blue whale. Source: Ocean Alliance.

may be transduced into biological information, scientific texts, and perhaps, eventually, conservation recommendations.

The team embarks on the second leg of the Sea of Cortez mission, south
to the San Ignacio Lagoon. After a twelve-hour drive to Bahía de la Paz,
they arrive, and their daunting goal is to work with less social whales,
such as humpbacks, fin, and blue whales. Ocean Alliance is in the Sea of
Cortez for one reason: developing the skill to collect about four liquid
droplets of whale blow per flight. This would be a fourfold increase over
the volume of biological material they collected the year before in Patagonia, Argentina, from southern right whales. In response, Kerr changes
his strategy, adjusting how he flies; where he positions the drone in relation to the whale, current, and wind; and whether he aggressively flies
through the whale blow or more passively allows the whale to come to
him. He needs to react to the elements of wind and waves, because they
have a tumultuous effect on the floating skiff from which he precariously
pilots. Molecular intimacies surface from this choreography of human
drive, drone technicity, and elemental perturbation.

Over three days, they fly forty-nine flights and collect about one drop of
whale blow each flight. To increase the volume of their samples, they attach

FIGURE 2.2. Drone's view of a blue whale before exhale collection. Source: Ocean Alliance.

more and larger petri dishes to their drone. Attuning to the atmosphere, they use the downdraft of the drone's propellers to push the blow onto the agar plates. On the final day, it is a success. They have captured eighty micrograms or four droplets of whale snot. Elements mediate—it is not just the drone's piloting but the operator's harmony with wind directions, the rolling buoyancy provided by the sea surface, and the oceanic drift of the animal. The interplay of the mediating atmospheres and the drone's technicity allows for this molecular gathering of vibrant mist (figure 2.1).

The second part of the Sea of Cortez mission is farther south of San Ignacio Lagoon. On the first day, after 160 kilometers of ocean travel, they do not see a single blow. They stop, turn off the motor, and listen. Nothing. Eventually, the team's patience is rewarded; they see a distant blow and race the drone toward the eruption. The air is an efficient material for mobility, speed, and agility as the drone flies forty-five miles an hour—much faster than their skiff. The source of the blow is a blue whale. Kerr has never been this close to the world's largest species. His nerves cause him to bungle his first flight. The whale is unfazed by the drone, as the sound waves of the propellers do not travel through the water (Christensen et al. 2016). But on Kerr's second try he swoops the drone in, collecting blow through a refracted rainbow of air saturated with misty exhale (figure 2.2).

Ocean Alliance's Rogan had this to say about the experience: "Better than all of this is something altogether more intangible. . . . This is such a simple, elemental [behavior]. The act of respiring, of filling the body with the oxygen all animals need to survive. But in such a vast, extraordinary and enigmatic animal . . . it is something more. It represents one of the few times we are able to gain glimpses into the lives of these enigmatic and often elusive animals, as they breach the surface of the ocean to breathe" (Rogan n.d.).

Rogan describes the "simple, elemental [behavior]" of whale breath as more than a quotidian act. He invokes "elemental" to refer to breathing as a primordial practice in an elemental milieu. Drone oceanography uncovers respiring organisms entangled with elemental forces. These elements contain, threaten, and sometimes destroy the work of drone oceanography. Elements also open possibilities: they challenge scientists and spur innovation; they keep their methodologies aloft, floating drones in the air and boats on the ocean surface; and they envelop their research collaborators, swathing whales in microbes.

The next year, Ocean Alliance is in the Keku Strait, Alaska, to trial a real-time artificial intelligence (AI) whale identification system designed by Intel. They are also there to create a segment for the National Geographic program *Earth Live*, a two-hour live wildlife documentary with camera crews simultaneously filming across six continents. In both projects, the atmosphere is a medium for the electromagnetic spectrum that entangles whales, software, and drones in a complex web of communication.

As they arrive in Alaska, Ocean Alliance and Intel are still hacking together the AI system, which they trained on images of humpback whales but have not yet used in the field. The AI is designed to automatically analyze fluke shape, compare it to an image library, and link that fluke to a known individual whale and its migratory and health records. This will allow them to track individual whales across their life span and help them understand how whales pick up, carry, and deposit microbes and pollutants. The system must be field-friendly, unintrusive, and run on a boat. They set up a small computer laboratory on the research vessel *Glacial Seal*. In a trial run they see blow, launch the drone, and fly above it, relaying video back to the ship. While the drone hovers, Intel's AI identifies—with 92 percent confidence—"Trumpeter," a humpback that has not been seen for twenty-three years (figure 2.3).

Such computer-aided processing is improving. Convoluted neural

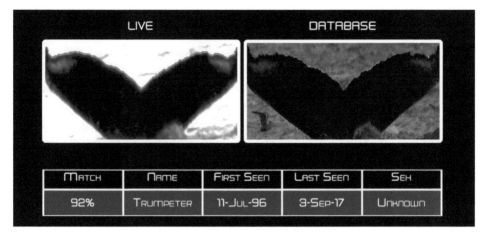

Live				
Match	**Name**	**First Seen**	**Last Seen**	**Sex**
92%	Trumpeter	11-Jul-96	3-Sep-17	Unknown

FIGURE 2.3. Whale fluke identification made using AI. Source: Bryn Keller 2018.

networks (CNNs) are a type of AI that builds on drone-collected photogrammetry, computer vision, and deep learning to analyze and make correlations between three-dimensional moving images. CNNs automate identification of blue, minke, and humpback whales with 98 percent accuracy (Gray et al. 2019, 1490). Images for training a CNN cannot come from satellites; they must be generated by drones because machine learning demands high resolution. Drones and CNNs mean increased speed and automation, which in turn aids identification: "If cetacean surveys are routinely conducted by UAS [unpiloted aerial systems] rather than ships and planes these automated capabilities will facilitate analysis and allow rapid management and ecological insights" (Gray et al. 2019, 1497–98). Enhancing speed and automation, Leroi-Gourhan ([1964] 1993) might argue, is the history of technology itself.

Another factor is at play that requires manual dexterity and fortune. The *Earth Live* documentary segment runs on an atmosphere and seascape that is saturated with network and visual technologies. Four cameras are required: the SnotBot (the name given by Ocean Alliance to their drone) swooping in and collecting a sample, another drone in the air filming it, a camera operated by a person on a boat filming these two drones, and a gyro-stabilized camera on an airplane filming everything below. The airplane doubles as a wireless router, relaying the footage to a satellite that beams it to a studio in New York City, where it is mixed with rushes collected from the five other live events around the world. This saturation or thickening of the atmosphere above the Keku Strait embodies transelemental mediation, the diffractions, transductions, and conversions of media that the elements afford (Fish 2022a).

In Keku Strait, Kerr is tethered by a cloud of networks to an eight-square-kilometer radius. If the live event is going to work and they are to be successful collecting whale snot on live television, the whales need to be within this small circumference. Tension builds in the minutes before the feed goes live from Kerr's drone. His immediate sight is to be magnified into visions for thousands of viewers. But no whale is visible. Kerr takes a chance. Without seeing a whale, he launches the drone and seconds later, as he describes it, "A BLOOMING WHALE SURFACED 500 FEET AWAY RIGHT IN FRONT OF ME—WHHHAAAATTTTTTT! IMPOSSIBLE!!!!! . . . Millions of people were snotted!!!!" (Kerr n.d.a). Kerr's exclamatory joy illustrates the affect of transelemental mediation, afforded by the drone's connectivity, mobility, and vision enabling a serendipitous choreography, leading to multispecies intimacy and an outburst of pleasure. Elementality suspended the drone's technicity that dilates witnessing, collapsing the space between a viewing public and a whale in a moment of planetary kismet.

CETACEAN SCIENCE

Whales are magnificent. They have extraordinary mental and communicative abilities, and strong, intergenerational social bonds. They live long, mature late, occupy the higher rungs of the food chain, and exist in pods. Sperm whales learned how to avoid whalers and taught each other the skill (Whitehead, Smith, and Rendell 2021). Whales are sentient and capable of joy and suffering (Nicol et al. 2020, 2). Humans have long seen and imagined whales—their prominence in our early philosophies and mythologies attests as much (Papadopoulos and Ruscillo 2002). They are some of the most photographed and scientifically investigated organisms in the world. From SeaWorld to Icelandic cruises, whale tourism is a multibillion-dollar industry. Cetology, or the scientific study of whales and other cetaceans, is a robust field because whales—massive mammals that spend a great deal of time near the surface of translucent ocean waters to breathe surface oxygen—are accessible to remote sensing and photographic documentation.

In modernity, whales rose to public consciousness with environmental activism of the 1970s, the first Earth Day, and the release of Payne's *Songs of the Humpback Whale*. Images of breaching, porpoising, and fluke-pounding whales became quintessential images of environmental-ism. These images and sounds are popular because they induce awe and sympathy—and because some species offer success stories. Several whale species have rebounded from industrial whaling, the result of whale oil becoming less valuable with the rise of petroleum in the 1970s, particularly after the International Whaling Commission agreed to a moratorium in 1986. The Pacific humpback whale is recovering, and populations of sperm whales are stable. In this manner, whale media mark the awareness of and resistance to the Blue Anthropocene—the demise of the life-sustaining ocean on a geological scale.

Whales embody the fragility associated with living in the Blue Anthropocene. Many cetaceans remain imperiled: the Atlantic gray whale and the Chinese river dolphin (*Lipotes vexillifer*) are extinct; the vaquita porpoise (*Phocoena sinus*) and the Māui dolphin (*Cephalorhynchus hectori maui*) are on the brink; a meager thirty North Pacific right whales are alive today; and blue whale populations have not significantly recovered since the global moratorium on their hunting—they remain at around 1 percent of prewhaling numbers. Yet whales also represent our ability and willingness to respond to the Blue Anthropocene.

Living whales matter. Whales are "sentinel species" of ocean sustainability; samples of their skin and breath inform us about the health of the ocean (Bossart 2011). Many whales regularly consume macroplastics (Alexiadou, Foskolos, and Frantzis 2019). Plastics being one class of evidence for the Blue Anthropocene, whale bodies alert us to the pervasiveness of our chemical regimes. The economic and ecological value of a single living whale is profound; their "ecological service" is estimated in the hundreds of millions of US dollars (Roman and McCarthy 2010). They absorb carbon through their food, and their feces provide nutrients for the growth of plankton deeper in the sea. This whale poop has been spotted by drones as it creates "pastures" for krill, small fish, and algae (figure 2.4) (Nealon 2019; Pancia 2019). When dead whales sink, the carbon in their bodies dives to the ocean depths and not into the atmosphere. One proposed form of carbon sequestration involves rehabilitating global whale populations to prewhaling levels. Bringing back blue whales, for instance, would be equivalent to preserving 43,000 hectares of temperate forest—the size of a megalopolis like Los Angeles. Rewilding the oceans with all whale populations would be equivalent to restoring 110,000 hectares of forest—the size of a medium-sized national park like Rocky Mountain National Park (Pershing et al. 2010). This research

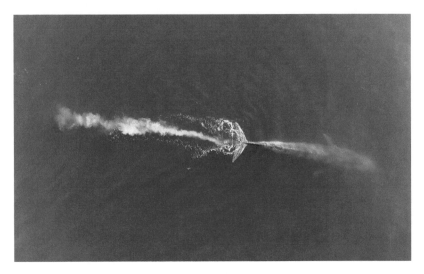

FIGURE 2.4. Blue whale fertilizing the sea around Australia as seen by drone.
Source: Ian Weisse (Pancia 2019).

"recasts whales as a means of re-naturalizing the air . . . whales as gardeners in the greenhouse" (Giggs 2020, 67). Cetology, or the study of whales, provides insights into how their lives, the ocean, the atmosphere, and our lives intertwine.

Traditionally, cetology gathered insights into whale populations, mating, migrations, and health from the leftovers of the whaling industry—skin, skeletons, stomach contents, and kill-location data (Burnett 2012). In the 1970s, cetologists began to shoot living whales with crossbow arrows that violently cut out samples of blubber. An atmospheric methodology, this crossbow biopsy is one iteration of a fascination with shooting things through the air and into whales in the name of science. Beginning in the 1920s, the British Discovery Investigations' expeditions into the Southern Ocean experimented with hitting whales with arrows, shotguns, and darts to track migration and reproduction. The idea was that a better understanding of whale populations and geography could be ascertained by linking the place the whale was tagged to where it was eventually slaughtered. A unique coterie of "sporting archers, colonial old hands, and creepy amateur medievalists" (Burnett 2012, 157) consulted on how to stick a whale with a dart. They settled on a twelve-gauge shotgun firing a contraption "as nonaerodynamic as could possibly be imagined" (159). Dozens of cetaceans were pursued on Discovery's first voyage, with zero whales tagged with the expedition's insignia (163). The purpose of this atavistic conservation science was to collect data in order to prolong the stocks for industrial whaling. It did not benefit the whales; it harmed them through pursuit and wounds fated for sepsis.

Having collected whale biopsies with crossbow and arrow, Kerr is able to compare drone oceanography to that violent process:

> I almost want to say I don't feel good. I've chased this animal down and whacked it with a crossbow. But when I am up in the air, literally flying down on these animals and I don't want to call it a day because I'm like, "Oh my God, look at that. Look how it swims through the water, look at its pectoral fins, the colors of the lights." . . . I would say it's actually a far more intimate act with the drone than it is with the crossbow. (Kerr, personal communication with author, December 2, 2020)

Contrasting methodologies beget divergent human-whale molecular intimacies and elemental attunements. For instance, the crossbow requires closer, more stressful contact between whales and people. Such "cross-species sensations are always mediated by power that leaves impressions, which leaves bodies imprinted and furrowed with consequences," feminist scholar Eva Hayward avers (2010, 592). Drones allow scientists a more intimate way to look, through a hovering-touching that creates copresence between biologist, drone, and whale, even at a much greater distance. Suddenly, what is touched is much more diffuse— exhaled mist instead of skin and blubber. The varieties of engaged material—billions of microbes—can be accessed from a pilot arriving relatively quietly from any direction.

Drone oceanography is both a continuation of the originary technicity of force from afar and a phase shift from the intrusive past of the crossbow plug. Cetologists began to develop new methods for procuring biological information beginning in the 1980s. Biologists collect samples such as the feces and skin that whales leave floating on the ocean surface. Dogs on boats in the Salish Sea of Washington State, for instance, are trained to use their sensitive olfactory systems to smell and direct researchers to data-rich whale poop (Donahue 2016). Divergent methodologies provide complementary snapshots of whale life: each collection praxis forges a singular entanglement, incurs a distinct level of responsibility, and creates an original human-whale intimacy.

Cobbling together such propulsion systems and projectiles exhibits the impulse to move and quickly eclipse the space between human and animal. Some of the first Paleolithic technologies—throwing sticks, nets, spears, projectile points, arrows, and atlatls—were designed to collapse space, exploit the atmosphere, and slow the pace of escaping prey. Hunting technologies gave way to technologies of enhanced seeing such as the sailor's telescope and binoculars that assist zoologists in witnessing and following animals from afar. Originary technicity enables us to trace the evolution of our tools back to our bodies. With it we can identify ontological constants across technologies.

Originary technicity refers to how bodies and technologies influence each other through time. Erect posture allows humans to see farther and appear larger, intimidating predators. Standing tall also gives humans a unique elevated position and enables movement for the hands to craft implements. An erect spine carries a brain whose cranial capacity blossomed and with it language and the ability to socially organize. The hand, freed from locomotion by two-legged walking, enables the grasping and use of tools. With these changes, the face opens and flattens, making more complex sounds possible. Here the free hands liberate speech, linking language and tools—which both exist exterior to and as extensions of our bodies. In this evolutionary process, thought itself is externalized in speech and spatialized through material culture.

Archaeologist Andre Leroi-Gourhan names this process of technicity "hominization," which expresses the human "urge to conquer space and time" ([1964] 1993, 26). This thinking began with Plato, cropped up in the work of technologist Ernst Kapp ([1877] 2018), was formalized by Leroi-Gourhan ([1964] 1993), was popularized by Marshall McLuhan (1964), influenced the work of philosophers Jacques Derrida ([1967] 1998) and Gilles Deleuze and Félix Guattari ([1972] 1977), and continues with the work of philosopher Bernard Stiegler (2018). Hominization is a liberation, exosomatization, or exteriorization of the body and senses out of the body through technology. In Stiegler's approach, "culture and society are not determined by technics but rather materialised through it" (Roberts 2012, 14). With technologies cocreating bodies, formulating the bodily extension by which the world is known and made, it is clear that "people are not just *in* the world, but *of* it" (Weston 2017, 16). Humans are "as and of" technicity. As technics that heighten human movement, seeing, and force from afar, drones concretize this coevolutionary, technological liberation of bodies in overcoming the limitations of space-time. The drone is a way of being in and of the sea and the atmosphere. It is a haptic technology of seascape epistemologies and aeolian anthropologies. The drone's relative technicity stretches the grasping hand and renders whale images into computational data. These acts of relative technicity fold whale into human bodies.

Kapp may have been the first modern to propose that technologies are extensions of human organs. For Kapp this goes in one of two directions. One route comes from Karl Marx, who, writing on Charles Darwin, celebrated how the evolutionist "directed attention to the history of *natural technology*, that is, the formation of the organs of plants and animals, which serve as the instrument of production for sustaining life" (Marx [1867] 1992, 493). This embedded technicity contrasts with Kapp's second technicity, or how technologies separate from the body and gain extrasomatic mobility. Philosopher Michel Serres (2007) names this "exodarwinism," and Stiegler (1998) calls it "epiphylogenesis"—evolution occurring from and outside of the body.

Leroi-Gourhan took a different approach to technicity. While Kapp argued that technicity was a matter of organ extension, for Leroi-Gourhan technicity was an externalization of motricity. The need to move, not corticalization or a growing brain, is the driving force for human evolution. Integrating the two perspectives are media studies scholars Ingrid Hoelzl and Rémi Marie, who theorize drones and other "machines of vision [as] based on two paradoxical moves: integration (in the human brain) and externalization (beyond the human body)" (2016, 72). To integration and externalization they add "posthuman vision," which is a "collaborative vision distributed across species and between machines/robots and humans/animals and any intermediary form" (73). The drone's relative technicity embodies these complementary potentials. It is an extension of the sense of sight, but also of movement. It is a tool whose capacity to see develops from a collaboration with a menagerie of nonhumans. As we saw with the work of Ocean Alliance, the drone's partial posthuman vision is *all too human*—contingent as it is on vagaries of human competency and elemental mediation. In its conservation work to preserve nautical thriving, it is integrated and externalized into the more-than-human whale world.

Before the drone, the camera was essential for cetology ceasing to study dead and suffering whales. Consider that, prior to the 1970s, orca researchers in the Puget Sound off the northwest coast of North America were not aware that some orcas (*Orcinus orca*) lived in the Salish Sea permanently and others were transient. From shore and boats, Dr. Michael

FIGURE 2.5. Parachute cetology in Patagonia, 1980s. Source: Ocean Alliance.

Bigg painstakingly photographed the Southern Resident killer whales and developed a visual key to identify individuals based on their black-and-white patterns or dorsal fin scars. Bigg discovered the local orcas and their small population (seventy-one in 1974, seventy-three in 2022), and began monitoring their population across generations. This allowed scientists to better understand and publicize how few Southern Resident orcas remained, which engendered conservation in nearby Seattle, Washington. Orcas are now icons of the risks and conservation responses possible in the Blue Anthropocene. With amplified speed, mobility, and posthuman vision, drones would have assisted Bigg immensely. As the partnership between Ocean Alliance and Intel showed, drones contribute their vertical vision to whale subjectification—the making of whale individuals.

Ocean Alliance's Roger Payne took a similar approach to individualizing whales when he conducted the longest investigation of whales in the world, focusing on the Southern Atlantic right whales of Patagonia, Chile, beginning in 1960. For Payne, Kerr, and their colleagues, photography was essential for identifying individual whales, and they wanted to elevate their cameras to obtain higher-resolution images. The first year they rented a helicopter, but it disturbed the whales. The coastline along Patagonia, where the right whales nurtured their young in shal-

FIGURE 2.6. Hydrogen balloon cetology in Patagonia, 1980s. Source: Ocean Alliance.

low bays, is lined with high cliffs. Payne and Kerr would run along them, following whales, stretching their arms to collect more revealing, less horizontal images. When they had the funds, they would rent airplanes. They tried more experimental techniques to obtain clearer vertical views. Parachutes and hydrogen-filled balloons lifted the scientists and cameras above nursing whales (figures 2.5 and 2.6). They "yearned to have a model aircraft that could carry a TV camera," Payne (n.d.) remembers. It would be decades before small, remote-controlled aircraft and video technology would be available.

Answering that yearning, Payne and Ocean Alliance pioneered drone cetology. As examined above, they developed drones to collect blow and that use AI to identify individual humpback whales. They have concocted thermal cameras on drones to gather data on whale body heat and health. But Ocean Alliance's most innovative application has been the use of drones to collect whale blow or exhale—a mission for which the drone's technicity of size, speed, mobility, and posthuman vision is ideally suited.

Field research with whales is intimate. Drone technicity allows closer contact or "data intimacy" between scientists and their research subjects (Calvillo and Garnett 2019). Data intimacy describes an increasing proximity, resolution, and actionability generated through intra-actions between sensors, elements, and bodies. The human scientist ventures into an often dangerous sea to track a massive animal and gather information to better understand its health. Confronting these risks requires epistemological and existential commitments. Such commitments to ethically know—and the responsibility that comes with that knowledge about sentience and its entropic fate—are crucial to this science. Ethical relationships are generated in acts of data intimacy and are dependent on the technologies used and their proximity to bodies. Proximity—how close you and your instruments come to the whales—matters. In the case of the SnotBot, proximity is such that the drones almost reach into the lungs of whales.

TOUCHING WHALE BREATH

Respiration involves lungs at work, inhaling and exhaling. "Breathing (in the elemental sense) is, first of all," writes philosopher David Kleinberg-Levin, "a primordial gesture of gathering" (1984, 124). Our elemental atmosphere is a chemical gathering of 78 percent nitrogen and 21 percent oxygen, with trace amounts of argon, carbon dioxide, noble gases, and water vapor. Whale lungs fortify a rich microbiome of bacteria and viruses. These microbes form an elemental atmosphere within the whale that is externalized during every exhale, then diversified upon inhale as the whale takes in suspended microbes. "Lungs have the largest surface area in the body in direct contact with the environment" (Dartnell and Ramsay 2005, 83), and whale lungs are the largest on Earth. Their sheer size makes them susceptible to inhaling pollutants and harmful bacteria; the lungs threaten to form a passageway from interior to exterior, dangerously exposing insides to outsides, increasing receptivity and transmission.

Drone oceanographers exploit this turning inside out. The sound and sight of breathing alert scientists to the whale's presence, the volume of its lungs, and the potential to collect whale exhalate. Blue whale respiration is the loudest breathing on Earth, and thus easiest to hear. For drone oceanographers, the sound is existential—a sudden awareness of a vast life—and a call to scientific action. The signal is a result of the immense size of their lungs, an expansion that occurred after they returned to the ocean fifty million years ago after a brief (geologically speaking) hiatus on land. Today, whales have carried their firmament-evolved lungs with them into the sea while their noses have migrated to the top of their heads. Respiration brings oxygen into blood circulation via the lungs.

Oxygen burns sucrose or triglycerides in the mitochondria of the cell, generating the cellular energy of adenosine triphosphate, or ATP. Wastewater and carbon dioxide are produced and exhaled. Respiring requires inspiration, the bringing in of this caustic force. The Latin *inspirare* means to "blow into, breathe upon." It was traditionally thought that an inhale was a divine power blown into living beings. The truth of that lore is now understood.

Breathing for a whale is not automatic as it is for humans; it is a premeditated act. If whale breathing is an intentional practice, then collecting whale breath by drone is also a technique of the atmosphere and of breathing, by both the human pilots and the whale. Whales breathe on drones with their internal winds. "Inspiration is wind becoming breath," in this case, and "expiration is breath becoming wind" (Ingold 2007, 31). Winds, technologies, and existential interventions mix in these acts of respiration. The drone navigates air and whale winds, flies inside warm exhaled mist, and, like the atmosphere itself, mediates ways of being and knowing, from whale to human.

Like a breathalyzer, the drone is a device that gathers something to measure from the breath (Peters 2017, 191). The goal of the whale drone, like the breathalyzer, is to understand health. Drones are like other breath-related oceanography technologies, drone oceanographer Karen Joyce writes: "In marine research, the advances offered by drones [are] arguably on par with the extent to which SCUBA diving revolutionised underwater research 70 years ago" (Joyce et al. 2018, 960). Mindfulness on breathing, key as it is to successful scuba diving, is an "embodied politics and ethics which can open us out to the ways we are intrinsically imbricated with other human and nonhuman others—particularly in this phase of a 'mass extinction' or 'biological annihilation,'" writes cultural geographer Chloe Asker (2020). Primordial, originary technicity and primeval breathing connect bodies and technologies in autochthonous ways.

As ventilating animals, we exhale and, as we do, microbes—bacteria, viruses, and perhaps fungi—exit our bodies and mix with the gases of the air. The air is an element, as is the water of the sea and the land of the firmament—each plays a part in suspending, holding, grasping, and mediating a transduction of molecular information. With respiration, "organisms continually disrupt any boundary between earth and sky, binding substance and medium together in forging their own growth and movement" (Ingold 2007, S19). Like inhaling is for whales, exhala-

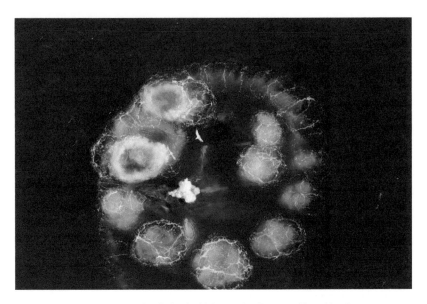

FIGURE 2.7. Two humpback whales bubble net feeding, as filmed by drone. Source: NOAA Fisheries / Allison Henry.

tion is also a curated practice. As Ocean Alliance and others have documented with drones, humpback whales blow circular "bubble nets" to contain and confuse prey before swimming through the gas column to take their quarry (Werth et al. 2019) (figure 2.7). Whale bubbles may also form acoustic barriers, creating enclosures that defend against the prying sonar waves of foreign whales (Peters 2017, 185). Exhalation, in the case of bubble net feeding, makes a light mist, a shuffle of breath that is warmed by the solar energy in the atmosphere or the metabolic actions of a sea animal. In this manner, the sun's radiation and "substances of the earth mingle and bind with the medium of the air" (Ingold 2010, S31). Drones in ocean research, too, mix the sea conditions and atmospheric possibilities, in the pursuit of data.

Land animals are not only of the hard earth but also of the atmosphere, which enters them through inhalation. Sea creatures "embody the climate of their environment through water" (Wainwright 2017, 343), but for the same reasons as their respiring land cousins, they are also of the air. Through breathing, swimming, hydrating, and bathing, human and nonhuman animals transgress porous elemental boundaries. Membranes, temporary as they may be, are films that separate the elements. As media studies scholar John Durham Peters (2015, 130) reminds us, membranes "are gates, one of the oldest of all media." Consider the gate that is the sea surface microlayer, also known as the neuston layer, or the top one millimeter of the ocean. It is the interface between the water and the atmosphere and is distinct from the layers immediately below it because it is richer in dissolved organic material. The neuston layer is the ocean that whales breathe in when porpoising. In the process, they forcefully inhale surface microbes deep into their lungs, where they may cause harm (Raverty et al. 2017, 2)—one of many concerns that oceanographers have about the fate of the Southern Resident orcas that swim in the polluted Puget Sound. In this manner, whales touch and are touched by the neuston layer, transducing external into internal microbes and chemicals, and externalizing their interiors to drone technologies. The actants in this interface blur the boundaries between elements; they are touching technologies and beings. The neuston layer mediates intimacies between microbes and whales, and whales and drones via exhaled breath.

Elementality, the concept of elemental mediation of messages and life, lessens the importance of things and instead emphasizes the substances between. The drone too is between; flying in the atmosphere between the sea and a pilot. Supporting this claim and building from the work of philosopher Jean-Luc Nancy, design theorist Benjamin Bratton writes that

> all touch is ultimately and finally touchless, that touchlessness is the basis of our intimacies. . . . Our most intimate encounters are mediated by sight and sounds, machines, bodily fluids, membranes, and prescribed behaviors. . . . His [Nancy's] most emphatic point is that we are touching and being touched constantly, and thus mediation is not a secondary condition of our embodiment, it *is* the condition of our embodiment. Instead of thinking of touch as that which is *im*mediate—without mediation—we understand instead that even as one experience may have more visceral tactility than another, that touching is always to some degree *touching at a distance*, and across a distance that is not empty but full of mediation. (2021, 73–74)

Thus, life is mediated—through the word, electromagnetic spectrum, DNA, drones, microbes, or the air itself. In this intraspecies becoming it is important, however, not to abandon the existentiality of being. These whales are in a struggle for survival, and the mediation of their insides can inform the blue governance of their survival.

Drones do not exactly touch whales; rather, they are enveloped in the whale's exhale. Relatives of the microbes that exist in the whale's lungs are transferred to the petri dishes on the drone. Although this is not flesh-on-flesh touch as it is commonly understood in human experience, originary technicity—the exodarwinian drive toward faster movement—enlightens drone-microbe-whale intra-actions. Drone pilots may be a kilometer away but see the whale more clearly than if they were directly touched by fluke waves. The drone touching the whale's blow is intimately connected to the drone's seeing. Drone vision makes whale intimacy. Two bodies, mediated by the elements, can exist in intimate relationship with one another yet remain distinct. The microbes are alive; the elements are platforms; and the agar-filled petri dishes stuck on the drone are only the most obvious media. Pilot eyes are focused through agile, remote-controlling fingers. Whales are harmlessly seen/touched, with helpful intentions. Given that human-cultivated pathogenic ecologies sicken whales, this multielemental mediation impels us to think about what is response-able—what can and should be done.

IN THE LAB

Ideally, after the sample is collected, the petri dishes will be opened in a laboratory and their organic materials sequenced and identified. But because of cost, time, labor, and other factors, little of the snot Ocean Alliance has collected has been analyzed until recently (Atkinson et al. 2021). This is unfortunate because the evidence could reveal the "microbiopolitics" of multispecies entanglements (Paxson 2008, 17). Following the work of science and technology scholar Helen Verran (2009), who implores scientists to reflect on the moral formations of their labor, I question the ethics of this analytical absence. To more fully complete the cycle of reciprocity that these scientific methods and instruments initiated, the whale's microbiome needs to be named and the fluid mobilities and narratives of polluted entanglements mapped. This is "science as activism," a more inclusive and ethical application of science's potential within life systems (Latimer and Miele 2013, 25).

In other contexts, drone-collected exhale shows that whales carry microbes from polluted seas (Apprill et al. 2017; Geoghegan et al. 2018; Vendl et al. 2020). Drones and other breath-based methods reveal that human-generated chemicals concentrate in whales (Pirotta et al. 2017). Chemicals, bacteria, and viruses in orcas in the Salish Sea, for instance, are entangled with humans, sewage, and environmental contaminants

(Raverty et al. 2017). Whale snot carries bacteria, viruses, and hormones that provide insights into whale health, their exposure to human-concentrated chemicals, and their position as vectors for the movement of microbiomes. The absence of the procedure to identify the microbiota in the drone-collected sample speaks to the limits of care and control of whale flourishing. Without this step, the caring control of blue governance remains on hold and the development of ocean/cultures is delayed.

In addition to chemicals and bacteria, marine biologist Vanessa Pirotta and her team found six novel virus species from five viral families in their drones' petri dishes (Geoghegan et al. 2018). Distinct from the "model ecologies" of symbiotic microbiota (Paxson and Helmreich 2014, 185), these "viral clouds" span seas, elements, and time, and "operate through infections and reassortments that are coincidental, responsive, opportunistic, and often nonrational" (Lowe 2010, 644). Aporic in their ever presence, viral clouds are "entangled and enfolded within a geography of more or less pathogenic landscapes" (Lorimer 2017, 547). Moving and irregular, with porous boundaries and ambiguous effects, whale viral clouds are examples of the kind of molecularization that science and technology scholar Michelle Murphy defines as "a vision of a world chaotically and dangerously interconnected by unpredictable viral exchanges" (2008, 697). Although unpredictable, the relation between the microbiome and the organism matters for survival and whether it makes a symbiotic or dysbiotic contribution to the ocean's health.

These viral clouds are expansive, surrounding and seeping inside other animals. Whales and seagulls simultaneously feed on the same small fish that whales consume. Feasting, flying, and defecating, seagulls share viruses with swimming whales, whose feeding practices require becoming engorged by water, fish, and seagull-defecated viruses. Through the "vast aerosol produced by whales . . . respiratory transmission may also play an important role in the movement of viruses in whales" (Geoghegan et al. 2018, 6). The size of their lungs and the largess of their exhale means that large pods of whales inoculate each other during blow netting and feeding.

Humpbacks are friendly not only to other humpbacks but also other species. They have been observed in mating-like activity with southern right whales, and swimming with fin, minke, blue, gray, and sperm whales. They play with bottlenose dolphins and protect seals from orcas (Pitman and Durban 2012). They may even save human lives. Whale biologist Nan Hauser's team flew drones and swam alongside humpback whales who defended her against a fifteen-foot tiger shark, blocking and pushing it away with their tails (Carbone 2018). Care requires proximity—a closeness that shares viruses between individuals and across species and technologies. These microbes travel between friendly whales and pods separated by vast seas to other species, and onto agar-coated petri dishes affixed to drones, which bring them back to laboratories. The whale microbiota that Ocean Alliance collects could draw our attention to material links and ethical relationships that might emerge from the drone's technicity. The ethics of this nonintrusive method involves care to minimize harm to whales and provide the data from which multispecies futures might be written.

Using the drone-derived microbiota data in a way that is meaningful to whales would be an "instrumentalization of DNA for social ends" (Waterton 2010, 160). The social preservation of long-standing whale communities would be possible if the "agency of non-human actors" were incorporated into "the marine spatial planning assemblage" (Boucquey et al. 2016, 5). The whale's rich microbiome—liberated at the moment of exhale, caught by an elementally suspended drone, sealed onboard a floating research vessel—has yet to inform marine conservation laws (Johnston 2019). The possibilities and threats of conservation technology and blue governmentality have yet to materialize. The intimacy of drone touching requires reciprocity, receiving, and giving. What humans give back remains to be seen.

This chapter has focused on the deployment of drones for the collection of whale snot. These operations reveal how technicity is made relative in intimacies between scientists, whales, and microbiota—all of which is mediated by unpredictable elements. The drone is in a tradition of technicity that includes crossbows and cameras, parachutes and hydrogen balloons and personifies the drive to collapse time and space. This trajectory moves toward improved grasping and grokking, as an extension of both the human and the posthuman body. In the ocean, care means conservation—achieved through the nonlinear and halting path of translating scientific discoveries into actionable policy. When the drone-collected microbiota are not analyzed, preliminary steps toward conservation remain untaken. Reciprocity is absent and flourishing flounders. In the next chapter, a more direct approach is taken to care for cetaceans with drones.

3

ELEMENTALITY
CONFRONTING WHALERS
THROUGH THE AIR AND
ON THE SEAS

COCAINE OF THE SEA OF CORTEZ

One late night in 2018, Sea Shepherd Conservation Society's mohawked pilot Jack Hutton flies a DJI Phantom—a small drone rigged with a heat-sensitive thermal camera—into a cloudless and starless night (figure 3.1). Leading a direct action within the campaign called Operation Milagro, Hutton stares at the screen of the drone's remote control, and using a handheld radio linked to personnel on the bridge of the cutter MY *Farley Mowat*, he receives a live feed of radio and radar information. He can use this data to triangulate the exact locations of the pangas or fishing skiffs in the area. Hutton is investigating illegal fisherfolk who are after a fat drum fish, the totoaba (*Totoaba macdonaldi*), in the Vaquita Refuge in the Biosphere Reserve, Sea of Cortez, Mexico. Called the "cocaine of the sea" (Ladkani 2019), the totoaba is prized for its expensive swim bladder, which Asian chefs value as a gelatinous soup ingredient; a kilogram of totoaba swim bladders sells for as much as $US80,000 in China (Rohrlich 2019). In the process of netting the totoaba, nearly extinct vaquita por-

FIGURE 3.1. Jack Hutton flying a drone at night in Operation Milagro. Film still from *Sea of Shadows* (Ladkani 2019).

poises (*Phocoena sinus*) are inadvertently caught and killed. They may be extinct by the time this book is published; in 2019 there were between six and twenty-two vaquita alive (Stevenson 2019).

A distorted voice on the radio announces, "So, Jack . . . the closest one is 0.9 miles. A little bit on the port side. And the one we saw five minutes ago is two miles away. On starboard." Hutton responds, "Copy that. Getting in the air now!" The air for Hutton is a medium for radio communication and the material that provides heft for the drone. It flies to document crimes and inform the authorities, who can arrest criminals. The goal is to stop illegal fishing so that the ocean can continue as a life-supporting medium for nonhuman mammals and others.

Hovering discreetly above the open-deck skiff, the drone's heat-sensitive optics provide a black-and-white vertical view of the warm bodies of five poachers aboard the craft. Hutton communicates to the ship's bridge: "I have a visual. I think they have a net. Yes, there is a net in there. They have totoaba on board." Thermographic footage reveals men hauling heavy nets by hand. Hutton shouts, "They are throwing the product overboard. Repeat, the product is overboard!" The skiff flees.

The drone is a tactical tool that provides situational awareness for Sea Shepherd. Hutton claims, "The military needs our help, because the Sea Shepherd drone can fly at five meters [above the sea] . . . and so I can confirm if there are swim bladders onboard. I can confirm if there are weapons onboard." Sea Shepherd radios the Mexican Armada as Hutton continues to track the skiff with the drone. The armada arrives, the skiff ignores a demand to stop, almost colliding with the armada vessel. The drone films the chase. Hutton calmly explains, "This is organized crime we are watching." The drone video is later integrated into Mexican television news broadcasts as well as the cinematic production of *Sea of Shadows* (Ladkani 2019), from which this tale comes.

Later in the campaign, a panga identified with the drone is seen unloading totoaba bladders onshore. The fishermen see the drone and quickly depart for safer, deeper water. The drone follows, and a figure stands on the skiff's bow. He fires a gun at the drone. It is hit. The video feed goes black. The drone loses power to its propellers and descends into the ocean. Hutton too collapses, saying, "the poachers don't want us looking at them, even if it means making use of automatic weapons. [This is] reaching a new level of violence" (Sea Shepherd Conservation Society 2017a, 2017b).

In their direct action, Sea Shepherd straddles criminality, vigilantism, and policing. By assisting the Mexican Armada to interdict poaching, Sea Shepherd's drones fly in an ambiguous space between militarism and conservation. Geographers claim that the drone's militarization of conservation has erosive consequences for conservationists, local people, and the species they are attempting to protect (Sandbrook 2015). It is true that enhanced surveillance can have negative consequences for humans. Local Mexicans, including those not poaching totoaba, are hostile toward the armada and Sea Shepherd, which complicate their fishing operations, legal or otherwise. Sea Shepherd is unfazed by the difficulties that its direct actions impose on criminals, even as Sea Shepherd's work is considered illegal by the local Mexican fishing mafia. Here drones contribute to the protection of nonhuman vitality by harming the vitality of human fisherfolk. This is one of the paradoxes of blue governmentality—how to balance care and control and manage life in fluid maritime space.

In Operation Milagro, Sea Shepherd's ideology of nature realism— that nature is distinct from culture and should be made more so—is externalized through the drone. The atmosphere mediates electromagnetic communication from drone to pilot, providing the lift for the drone as it tracks the skiff, and fueling the bodies of the Sea Shepherd volunteers, totoaba, and vaquita. Some poachers are disciplined; most escape into the dark night and the smooth sea. Blue governance stalls.

For over forty years, Sea Shepherd has campaigned in all the world's oceans to stop whaling, shark finning, gillnetting, seal culling, and other aspects of industrialized fishing that they deem unsustainable. Sea Shepherd deploys a legion of atmospheric technologies such as helicopters and drones, and a fleet of oceangoing vessels. Atmospheres lift their drones and helicopters, and their vessels float, sail, and power through the oceans. These elemental vehicles carry people and technologies to stop violations of conservation laws and principles of nature realism. The ocean, wind, ice, waves, rain, snow, and storms are constant sources of elemental pressure. The atmosphere and ocean are channels for movement in convulsive seas and direct actions to conserve marine species.

In this chapter I investigate numerous primary and secondary documents, including the autobiography of Sea Shepherd's founder, Captain Paul Watson (1981), direct-action manuals (Watson 2020), journalistic accounts (Heller 2007; Khatchadourian 2007; Scarce 1990), legal studies (Nagtzaam and Guilfoyle 2018), cultural interpretations (Robé 2015), long-running television programs such as *Whale Wars* (airing from 2008 to 2015), documentaries like *Sea of Shadows* (Ladkani 2019), numerous Sea Shepherd blogs, several interviews, site visits, and other digital and video artifacts as evidence of elemental mediations and technical tendencies.

Sea Shepherd's nature realism is biocentric and aggressively opposes destructive human agency, capitalism, and industrialism (Naess 1973). It is an embodied philosophy and a practice-oriented ideology that resonates with anarchism and deep ecology; anarchist Murray Bookchin (1971) emphasizes the physicality of anarchism to break and remake society. Politics and lifestyle merge into prefiguratism, a politics defined by living the future that radical environmentalists envision through anticonsumerism and simplified lifestyles. This nature realism is "vitalist, holist, organicist" (Ingalsbee 1996, 268). It is a praxis, an action philosophy that physically defends autopoietic systems of self-regulating and self-reproducing flows of information, energy, and vitality (Luhmann 1986).

Conservation technicity manifests extrinsically as "liberations" or externalizations of internal, behavorial, or cultural frameworks. In the case of Sea Shepherd, their nature realism is an "interior milieu" (*milieu intérieur*; Leroi-Gourhan [1964] 1993). An interior milieu conditions a "technical tendency" (*la tendance technique*), a disposition toward local and cultural materialities. Sea Shepherd's interior milieu is a nature realism that is regulated by a critical realist faith in technoscientific tools, methods, and findings. Sea Shepherd's interior milieu is biocentric; "our clients are whales," claims Sea Shepherd managing director Jeff Hansen (2020). Watson's (1981) biography, his direct-action manual (Watson 2020), and his three "laws of ecology" advance this interior milieu for biodiversity and multispecies interdependence and against the calamity of anthropocentrism (Watson 2019). A rigorous interview process ensures that these principles are shared by volunteers on Sea Shepherd crafts.

In this formation, the behavioral or interior milieu bleeds outward, through the physical or technical milieu, toward a geographical, environmental, or exterior milieu (*milieu extérieur*). Leroi-Gourhan influenced philosopher Georges Canguilhem (1965), who identified how "the behavioral milieu coincides with the geographic milieu, the geographic milieu with the physical milieu" (Canguilhem and Savage 2001, 17). This interior milieu is liberated and exteriorized into the atmosphere and over the ocean through conservation technologies. Ships, helicopters, and drones are three examples of liberations from this interior milieu to an exterior milieu of endangered whales and porpoises, atmospheres and oceans, weather and ice, whalers and poachers. The drone's design, both at-hand and as an atmospheric thing, is the expression of an evolutionary force that connects the hand and eye to technologies and through itself to an exterior milieu of elements, criminals, and cetaceans.

In nonactivist contexts, the drone's engagement with the elements is clear. Drones are engineered to collect water and sediment (Di Stefano et al. 2018), ocean temperature (Inoue and Curry 2004), ocean aerosols (Corrigan et al. 2008), salt from the surface of the sea (McIntyre and Gasiewski 2007), chemicals and particulates from the air, and, as discussed in chapter 2, whale breath vapor. Drones monitor their own elementality, gathering from the air evidence of fatal amounts of carbon monoxide, sulfur dioxide, and lead (Bolla et al. 2018). Atmospheric drones cross the elemental membrane in the pursuit of pollution, landing on water to test for toxic cyanobacteria (Coxworth 2019). Drones sail the sea and dive below the surface. They have even parented other sensors, dropping sensors in habitats to monitor for the presence of wildlife (Di Stefano et al. 2018). This chapter is an empirical investigation into the technical tendencies of elemental technologies such as the drone, as well as the helicopters and oceangoing vessels that attempt to enact governance over the oceans from the air.

Sea Shepherd's interior milieu is exteriorized through direct action—the slowing or stopping of industrialized extraction and capitalist transaction through physical interventions, blockages, disturbances, and theatrics (Graeber 2009). Technologies figure prominently in direct actions. Consider ecotage (or monkeywrenching)—a mode of civil disobedience that applies bodies, physical force, and tools to damage extractive technologies and impede extractivism. Direct action is confrontational; it situates the state or corporations as antagonistic exploiters of life itself. The tactics of direct action are deployed to combat environmental or illegal destruction, or in the defense of other species. "Sometimes, to dramatize a point . . . it is necessary to perform a violent act. But such violence must never be directed against a living thing. Against property, yes, but never against a life," argues Watson (Nagtzaam 2014, 632). For many radical environmentalists, direct action is voluntary work inspired by devotion to the guardianship of biological diversity. For others, it personifies an expansive notion of enlightened, transpersonal self-defense. Radical environmentalists who espouse deep naturalist biocentrism and physically intervene in ecological destruction use all means available to stop extinction (Scarce 1990). They embody a relative technicity; moving faster on the ocean, over land, and through the air to quickly document marine pollution, disrupt poaching, and stop illegal fishing. Sea Shepherd's attempt at blue governmentality through navigating conservation technologies is executed in a weather-world (Ingold 2010). Here the elements mediate interactions by connecting interior aspiration to exterior environment, and blue governmentality attempts to care for and control ocean biopower.

ELEMENTAL MEDIA

The ocean's wetness, fluidity, and instability and the atmosphere's gaseousness are elemental qualities that afford and complicate lifeways. The ocean's "more-than-wet ontologies" emerge from a coconstructive relationship with the land, wind, and the various states of the sea—flowing, frozen, fog, or mist (Peters and Steinberg 2019). The ocean and atmosphere enhance and forbid mobility for humans and their technologies. Elements weather the world, exposing and eroding humans' attempts at governance (Engelmann and McCormack 2021). For this reason, it is necessary to think elementally about governmentality. It is through these elements, accessed by technologies, that humans attempt to manage the world.

Elements mediate. As forces that enable crossing, they arbitrate contact between spaces, subjectivities, and species (Adey 2015). This is so because the elements exist, as communication scholar John Durham

Peters writes, "between sea, earth, and sky" (2015, 12). Elementality describes substances between, through which connections flow. Media studies scholar Nicole Starosielski contrasts the technocentric ontological theory of media of Marshall McLuhan and Friedrich Kittler against an antiontological media studies undergoing an expansive "turn to nature" and "turn to water" (Starosielski 2021, 306). Elementality describes the functions of this rewilding media studies that "extends the range of substrates that can convey meaning" (306). The materiality and signals of communication technology are themselves mediated by the elements.

For Sea Shepherd, the electromagnetic spectrum in the atmosphere makes possible communication between ocean-cruising vessels, drones, and helicopters. The air offers a thick materiality on which their drones and helicopters fly. The ocean is also a material that floats their multiton ships. The whales and porpoises they try to protect glide through this physicality. Most essentially, the ocean is a medium for cellular respiration for both human and marine mammals. Watson (2015) is fond of reminding listeners, "if the ocean dies, we all die." He is doubly correct: the oceans are a source of immense dietary enrichment, and over half of the world's oxygen comes from phytoplankton. In this manner, the elements mediate and modulate how Sea Shepherd's living bodies and interior milieux are externalized through the drone's relative technicity.

In their long hauls on ships in pursuit of poachers, the activists acclimatize to what feminist water philosopher Astrid Neimanis (2017, 54) calls "hydro-logics," which include the ocean as a system for gestation (e.g., its life-supporting qualities), communication (e.g., its articulation of sound), movement (e.g., its carrying of vessels), archiving (e.g., its storage of floating and suspended artifacts), sculpting (e.g., its terraforming through wave action), and differentiating (e.g., its action as a dissolver of solids). Hydro-logic informs a "sea epistemology" (Ingersoll 2016), which blends with an "aeolian anthropology"—an attunement to the affordances and disruptions of the air and how it scales space and time, enhancing movement and vision (Choy and Zee 2015; Kenner, Mirzaei, and Spackman 2019, 154; Peterson 2017). Analytical categories and experiential recursions are rendered soluble by political passion and wave action: "Both the ocean *and* our categories for understanding it complicate efforts at regulating extraction, governing mobility, or fixing histories" (Lehman, Steinberg, and Johnson 2021, 17). Elemental logics, sensibilities, and attunements transduce, conflate, and confound, creating a synesthesia of windswept weather-worlds and computer vision.

FIGURE 3.2. Sea Shepherd's cutters MY *Sharpie*, MY *Farley Mowat*, and MY *John Paul DeJoria*, filmed by drone. Source: Sea Shepherd Conservation Society 2018.

A brief analysis of Sea Shepherd's ships provides an example of how their interior milieu of nature realism connects to and moves through an exterior milieu of elements. Sea Shepherd's fleet of eleven ships includes the MY *Bob Barker*, an 800-ton ice-classed vessel, and several 170-ton decommissioned US Coast Guard patrol cutters (figure 3.2). In their mobility and more-than-humanness, these ships are exteriorizations of Sea Shepherd's action-oriented interior milieu (Hasty and Peters 2012, 671). Sea Shepherd leverages the pan-Cartesian maneuverability that the ocean provides to ram and sink boats engaged in illicit action, obstruct vessels from entering ports, destroy whale-meat processing facilities, and collect and disable drift nets (Nagtzaam and Guilfoyle 2018, 364). The ocean endows these direct actions with buoyancy and fluid mobility, while at times inhibiting the vigilante justice that these ships were deployed to enforce. Elements funnel the movement and vision of originary technicity into context-specific relative formations. The communication and connection of intimacy are made possible by these atmospheric and oceanic media.

From their first concerted direct action in June 1975 aboard the *Phyllis Cormack*, pursuing Russian whalers in the Pacific Ocean fifty miles west of Eureka, California, Watson and Greenpeace, the organization he cofounded, were attuned to radio waves traveling through atmospheric media. Their vessel was loaded with wave-form transforming instruments for elemental and interspecies communication with whales: "a balky Tcherepnin synthesizer, a quadraphonic underwater sound system, and a full hydrophone array—the better to facilitate the trippy, spatialized, symphonic encounter with these alternative intelligences of the deep sea" (Burnett 2012, 642). But while the activists were interested in hearing from whales, they really wanted to get wind of Russian or Japanese trawlers. With speakers of both languages onboard, they would bend their "ears to the scratch and squawk of the boat's old radio for some hint of a Russian accent" that might inform them of the whalers' presence and hopefully, eventually, the bearings that Greenpeace could follow in disruptive pursuit (Watson 1981, 20).

They not only listened to the air but populated it with their own sounds of whale songs. From on-deck loudspeakers, they blasted the Soviet whaling ship the *Dalniy Vostok*, once they found it, with the songs of humpback whales collected by Ocean Alliance's Roger Payne, whose research was discussed in the previous chapter. In this campaign, the air was not only an opportunity for hearing, seeing, and sounding—its transparency also made the activists vulnerable to being seen. Surveilling from an elevated vantage point, the Canadian Air Force relayed the antiwhaling activists' location to the Soviet embassy and eventually to the whaling ships. With this information, the whalers avoided Greenpeace, exposure, and further conflict.

In one encounter, these early founders of Greenpeace used small, rigid inflatable boats (RIBs) to track and harass the *Dalniy Vostok* by positioning themselves between the Soviet ship and sperm whales. The Soviets fired a harpoon. It whizzed past the ears of Greenpeace's Robert Hunter and George Korotva, and then an acoustic thud reverberated from the whale's body. The event was collected on audio- and video-recording devices by Greenpeace for transmission across television airways and film channels. From these first encounters, ocean activists began using and being abused by the elements.

The ship is a conservation technology that allows activists to situate their bodies directly within a conflict, widen their vision, and project the images they generate into other human worlds. "For centuries the ship was to geography and geographers what the telescope was to astronomy and astronomers; a means of extending the 'vision' of their science," write geographers William Hasty and Kimberley Peters (2012, 662). Today, Sea Shepherd's ships carry volunteers and their technologies of intervention with an acceptable level of risk. They sail them to the Southern Ocean to oppose whaling, and to the Sea of Cortez to stop the slaying of porpoises. As such, the ship is an existential technology for both the endangered species and the volunteers. It preserves the lives of Sea Shepherd volunteers in the whalers' counterattacks and shields them in gales among "growlers"—hidden, hull-piercing icebergs.

When sailing through these ice fields in pursuit of whalers, "you are no longer in control of your vessel. You have relinquished control to Mother Nature," states Sea Shepherd helicopter pilot Chris Aultman (*Whale Wars* 2009a). In one tense encounter in 2009, icebergs pushed against the hull of the *Steve Irwin*, flexing the walls and causing small leaks. Sea Shepherd volunteers had to sit deep in the frigid, creaking hull under the waterline, near the pulsating bulge. They were ready to help stop the leak and be the first to sound the evacuation alarm (*Whale Wars* 2009a). Their ships and bodies were permeable, "integrated within, and subject to, a global system: one that combines the air we breathe, the weather we feel, the pulses and waves of the electromagnetic spectrum that subtends and enables technologies, old and new," writes technology scholar Frances Dyson (in McCormack 2018, 210–11). The ship is an envelope around which waves of media and material reverberate.

Focused as it is on the originary technicity of force, elemental activism controls, diverts, suspends, and floats itself and other actants. One gruesome elemental methodology involves using nail guns to shut the scuppers from which whale blood flows from flensing decks, so that it does not flow overboard and thereby disturb Japanese whalers in the Southern Ocean (Khatchadourian 2007). Sea Shepherd also built concrete and rebar contraptions called "net rippers," which they deposited on the ocean floor in the Grand Banks in the North Atlantic Ocean. This direct-action technique destroys nets that trawl the seabed—much like Earth First! forest defenders who warn when they have spiked trees, a technique of hammering long nails into trees. If a chainsaw hits a nail while felling a tree, it will dangerously buck or bounce. Loggers will not cut if trees are spiked. In this fashion, Sea Shepherd announces the sinking of net rippers but not their actual location (Khatchadourian 2007, 6). No fishing vessel could drop nets for fear of destroying their expensive tackle. Sea Shepherd seeks to democratize this underwater activism, inviting supporters to build their own "hedgehogs" and "aquatic caltrops" and drop them in overfished seas (*Earth First!* 2005). Positioning activists to block the flow of surface-sloshing blood and impede the floatation and fatality of nets, the ship allows Sea Shepherd to project their interior milieu into an exterior milieu of illegal fishing and whaling in savage seas. In the section that follows, Sea Shepherd's interior milieu meets a particular exterior milieu, that of the atmosphere, and a distinctive elemental vehicle, the drone.

ATMOSPHERIC ACTIVISM

As they exploit the elements and are liberated from the interior milieus of deep naturalists, drones are extensions of at least three technical tendencies: situational awareness, documentations of illegality, and visual spectacle. All three tendencies can be found in Operation Milagro. The drone's technical abilities in this campaign included the production of situational awareness that informed the crew of the Mexican Navy or Armada's interdictions. It also enabled documentation of poaching, leading to the generation of ecological spectacle for the award-winning film *Sea of Shadows* (Ladkani 2019). Also attesting to this trifecta is the work of the director of Sea Shepherd Scandinavia, Lucas Erichsen, a veteran of nine campaigns. He flew drones to assist Sea Shepherd in preparing to rescue whale sharks (*Rhincodon typus*) from nets, connect diesel fuel spills with illegally leaking crafts, and collect gruesome images of long-finned pilot whale (*Globicephala melas*) offal dump sites (Erichsen, personal communication with author, February 2, 2020). Sea Shepherd's drones assist them in effectively planning missions, recording illicit fishing, finding criminals, and producing disturbing images of butchered marine species. Sea Shepherd's drone work combines the monitoring aspects of drone oceanography with the interventionist applications of drone activism. The objective is blue governmentality—a caring control through the mechanics of relative technicity and from an interior milieu of nature realism—as mediated by elemental contingencies. These technical tendencies strain to perform blue governmentality across the sea and from the air. Paradoxically, Sea Shepherd's attempt at control is in tension with their goal: oceans devoid of human arbitration.

Sea Shepherd has been in a decade-long battle with Japanese government–supported whaling. The Japanese Institute of Cetacean Research (ICR) claims a scientific exemption from the 1982 (implemented in 1986) International Whaling Commission's moratorium. Ships, helicopters, drones, and an assemblage of sensors—binoculars, radar, automatic identification system, GPS (global positioning system), barometers, and the internet's data streams—contribute to Sea Shepherd's sensing technologies as they work the airwaves. From 2011 to 2012, in Operation Zero Tolerance in the Southern Ocean, Sea Shepherd deployed four ships, a helicopter, eight RIBs, and three drones to stop the ICR's slaughter of over one hundred fin whales (*Balaenoptera physalus*). Expressed in the terms of relative technicity, they hoped that these tools would provide "the necessary combination of speed, range, strength, and communicative abilities to achieve this zero tolerance objective" (Sea Shepherd Conservation Society 2012). Increasing speed, range, strength, and communication are precisely the primordialities that Leroi-Gourhan ([1964] 1993) theorized for all technologies.

As the drones' range, battery power, and durability increased through the years—while fuel costs for piloting their Sikorsky S-300 and MD 500 helicopters remained astronomical—Sea Shepherd began using more drones and fewer helicopters to find whalers. Erichsen told me, "every five minutes the helicopter uses basically the amount of fuel that it would cost to buy a new drone" (personal communication 2020). The drone's cost-effective atmospheric movement increases situational awareness, a state of consciousness defined in the drone literature as "the perception of the *elements* in the environment within a volume of time and space" (Drury, Riek, and Rackliffe 2006, 88, italics added). The drone, like the helicopter it replaces for short missions, exploits the lifting materiality provided by the atmosphere's thick gases to furnish Sea Shepherd with an expanded, vertical, adaptable, and mobile viewpoint. This information helped Sea Shepherd coordinate attacks on Japanese harpoon ships and the whale-meat processing vessel—the eight-thousand-ton colossus the *Nisshin Maru.*

The drone is also effective in identifying and documenting crime. In 2019, a Sea Shepherd drone helped find one of the most wanted poaching vessels in West Africa, the STS-50 (Gray 2020). In 2017, inside the Coiba National Park, off the Pacific coast of Panama, a Sea Shepherd drone documented their MV *John Paul DeJoria* being rammed by an aggressive fishing ship (Sea Shepherd UK 2017). In these actions, the atmosphere's material lift but also its potential as "elemental infrastructure" (McCormack 2016)—electromagnetic waves that enable the transmission of images of felonious deeds—make it an effective tool. Obviously, the atmosphere is also a medium for abetting the lungs of Sea Shepherd volunteers and marine mammals. In this elemental activism, the drone's technical tendency toward policing does more than targeting, maiming, and killing from above, which some military drone critics would have us believe are drones' main purposes (Chamayou 2015). An incommensurable biopolitical magnitude is amplified by the ocean drone. Liberated from an interior milieu of nature realism, spread into an exterior milieu of illegal fishing by direct action, the drone's technical tendencies can be deployed toward blue governmentality, a state of ocean-specific surveillance for existential containment moderated by sea swell and storms.

While merely witnessing atrocity is not Sea Shepherd's first priority, impressive spectacles originate from the drone's mobility, receptivity to electromagnetic radiation for direction, and recording of threatened and often magnificent species. As a student of media scholar Marshall McLuhan in Vancouver in the 1960s, Watson (2019) understands the importance of spectacle. This spectacle, however, is not the cynicism of Guy Debord's (1994) Marxism or Jean Baudrillard's (1983) "hyperreal" constructivism, but rather a fugitive propaganda from a marginalized milieu. Direct-action nature realism is a counterhegemonic insurgency that is fundamentally contradictory to capitalist extraction—while rarely threatening it. This "radical media" (Downing 2001) makes representations that galvanize support for political operations. Drone spectacle is a will to visibility "at the intersection of the two domains of aesthetics (relations of perception) and politics (relations of power)" (Brighenti 2007, 324). These outsider aesthetics are generated by exploiting the material lift of atmospheric gases and the communicative capacities of electromagnetic media. Through this network of atmospheric media, broadened into the elementless vacuum of space to bounce from communication satellites, and back down to personal and collective screens, these images hail attention, outrage, and organization-sustaining donations. In this manner, the drone is a medium for life—it contributes to Sea Shepherd's existence and gives an advantage to activism on behalf of multispecies prospering.

The potentials of blue governmentality to protect marine life shift in response to elemental affordances, human capabilities, and access to interventional technologies. In contexts not on the sea, divergent milieus, technologies, and elements present contrasting complications for governmentality. Toward the goal of stretching the application of this chapter's concepts beyond drones and the sea, the discussion that follows expands atmospheric activism, elementality, and technology into their transduced formulas. Transductions within multispecies assemblages refer to turbulent translations with animal, technological, and elemental distortions (Helmreich 2007, 631). Applying transductive pressure to key concepts results in *atmoactivism*, which highlights the diversity of atmospheric mediations; *transelementality*, or a translational force across multiple elements; and *transtechnologies*, or technological multiplicities that mingle, cross, and diverge. Thinking transductively about milieus, technologies, and the elements exposes the challenges of governmentality with drones or other atmospheric technologies, in the air or city, on the sea or land.

ATMOACTIVISM

Sea Shepherd has a long history with atmoactivism, or activism with the atmosphere. Watson chartered an airplane in 1982 and dropped light bulbs filled with blood-colored paint on a Soviet whaling vessel (Khatchadourian 2007, 9–10). He fired rotten pie filling through the air at opponents (Khatchadourian 2007, 9). Mixing water and air, Sea Shepherd shot water cannons at harpoon vessels, blasted them with paint guns, and launched foul-smelling butyric acid at whalers (*Whale Wars* 2009b). Watson claims to have invented "the first organic biodegradable nontoxic delivery of chemical warfare ever when we hit them [the ICR] with butyric bombs and methylcellulose" (*Whale Wars* 2009b)—methylcellulose, when mixed with water, makes decks slippery and dangerous for work. Deploying such tactics against ICR whalers, Sea Shepherd's helicopter buzzed, harassed, and surveilled the harpoon ships. This direct action exploits the air's potential for unblocked mobility to generate havoc for Sea Shepherd's enemies while expanding the notion of atmoactivism.

When accessible, the helicopter is an impactful technology. Like the drone, the helicopter's elevated vantage point enhances situational awareness and the collection of evidence of illegality and spectacular images. Former Sea Shepherd pilot Chris Aultman reported that "the helicopter is the quickest, most decisive way to verify a position or search for something" (*Whale Wars* 2009b). In one mission against ICR whalers in 2011, Aultman launched and conducted transects all day and in every direction, returning only for fuel. He found the *Nisshin Maru*, the whale-meat-processing behemoth and the ultimate target for their goals of disrupting whaling. To his horror, he saw and recorded the carcass of a fin whale being cranked onto the flensing deck. Killing a whale is an orgy of atmospheric violence. The harpoon gun fires an explosive-tipped spearpoint, which makes skin and organs explode from the whale (*Whale Wars* 2009a). A rifle finalizes the killing. Aultman's gruesome image was relayed around the world via cable, satellite, and internet video. The helicopter is an atmospheric technology capable of elevating video cameras, generating spectacle, and extending interior into exterior milieus—and, in a manner typical of blue governmentality, with mixed results.

Early in its history, helicopters transported Sea Shepherd to the ice floes to stop harp seal (*Pagophilus groenlandicus*) hunting in northeastern Canada. The harp seal is a mammal that adapted to swimming, fishing, and living in the sea. Like all seals, they give birth on solid surfaces, exposing vulnerable young to predation by humans who can navigate the slowly shifting labyrinth of Labrador icebergs. Three Canadian and five Norwegian vessels were harvesting the ice-white skins of six-week-old seal pups in March 1976. Sea Shepherd utilized two Bell Jetranger helicopters to find the sealing fleet, deliver supplies, bring the activists into direct confrontation with the sealers, and save as many seals as possible. The helicopters were expensive but provided the tactical advantage that the activists needed to find and travel to the sealers and their prey. The helicopter offered a vertical viewpoint on the carnage: "Great smears of blood . . . in the diamond-hard ice" (Watson 1981, 92)—the result of 13,600 seals killed in the first day.

Helicopters were essential in ferrying the activists from the camp to the sealing grounds. Recognizing this dependency, the Royal Canadian Mounties charged the pilots with violating the Seal Protection Act and confiscated the helicopters (Watson 1981, 99). Meanwhile, Canadian Fisheries Ministers—Royal Canadian Mounted Police tasked with managing the sealing—continued to fly sorties to the action (98). The helicopter was not the only atmospheric tactic used by Sea Shepherd in this early action. They sprayed the white seal pups with aerosol-delivered red dye, destroying the pelts' value. Sea Shepherd made public their plan to spray from the helicopter (80). Upon learning of this plan, the Canadian government prohibited helicopters from flying low over seals (83). Control over airspace was imposed by state law and the state's legal monopoly on violent interdiction.

The 1977 campaign ended with Watson handcuffing himself to the sealers' winch line and volunteering for elemental torture. He was dragged across the ice, dipped into the freezing ocean, hung in the air from the bow of the seal-skinning vessel, beaten, force-fed seal blubber, witnessed seals being skinned alive, and was eventually sent to the police on a helicopter (Watson 1981, 145–50). At the campaign against sealing in 1979, one police helicopter nearly landed on Watson. He claims the pilot was

FIGURE 3.3. Atmoactivism in the form of protest gas deployed in Iceland by former Sea Shepherd volunteers. The white belly of a deceased fin whale can be seen floating beyond the protesters' smoke and kayak. Source: Pierschel 2017.

"trying to push me down and off the ice" (196). Watson crouched and threatened the helicopter's whirring propellers with his walking staff. Exploiting the atmosphere—afforded by technologies, regulated by law, and exposed to elemental precarities—remains dangerous activism.

Focusing on fin whaling in Iceland, former Sea Shepherd volunteer Arne Feuerhahn started Hard to Port, a "whaler watching" organization that combines two atmospheric technologies: drones to document the butchering of cetaceans and orange smoke grenades to highlight the carcasses (figure 3.3). In this manner, Sea Shepherd and allies expand atmoactivism. To drones and helicopters, they add bloodred light-bulb bombs, pie filling, and gas dispersions to generate confusion and theatricality.

Atmoactivism should also be considered terrestrially. Much has been written about the constitutive role of networked technology in protest formation and message dissemination (Srinivasan and Fish 2011, 2017). Less understood are the non-networked "other media" alive in the atmosphere in protest actions (Feigenbaum 2014). Less technological and more "atmospheric" in the affective sense of the word (Gandy 2017), the air itself can be mobilized for political gain. Music, chants, and speeches have long been carried by the wind and are integral to marches. During direct actions, atmospheric gases enable bubbles and balloons to register tear-gas direction, and lasers and fireworks to disorient police forces (CrimethInc. 2020). Portable diesel-powered leaf blowers were used to disperse tear gas during the 2020 protests in support of democracy in Hong Kong and in advocacy for racial justice in Portland, Oregon (figure 3.4). The leaf blowers pushed the caustic gas back on the police and

FIGURE 3.4. Leaf blower (left) deployed as an atmoactivist technology in Portland, Oregon, 2020. Source: Jon Gerberg, Whitney Leaming / *Washington Post*, 2020, reproduced in Lang 2020.

federal agents who dispensed it (Lang 2020). In Portland, police used leaf blowers to blow back the tear gas they had initially unleashed. Tear gas is not only atmospherically ephemeral; its political impact is blowing in the wind.

These multiple and sometimes dueling efforts to politicize or indeed weaponize an element with swirling and shifting forces create the multidirectional, multitechnological, and transelemental atmoterrorism that philosopher Peter Sloterdijk (2009) predicted. In this manner, atmoactivism is like other aspirations, political or otherwise, to liberate human mobility into the elements and distribute intentionality across space. These experiments exhibit the political bifurcations of elemental liberation. As a vigilante policing force, Sea Shepherd's drones shoot the gap between the life-defending liberation of atmoactivism and atmoterrorism, the dangerous securitization of and by the elements.

TRANSELEMENTALITY

As climate science has all too graphically proven, the elements are intertwined. A warmer ocean puts more moisture in the atmosphere, resulting in more rain, hurricanes, and intense storms. The ocean absorbs atmospheric heat. As substances that seep, flow, fill, and find their way into open volumes, the air and water leak into each other. Elements trouble separation; they "co-saturate each other" (Jue 2020, 15). In this manner, drones as flown by humans—who are usually rocking on boats in ocean swells, monitoring marine looting from the air—force us to consider transelementality, the mediations afforded by transductions, or system-dependent translations, across elements. Previously I've discussed the drone's transelemental qualities, and here I return to Sea Shepherd's ship as a vehicle for oceanic/atmospheric transelementality to demonstrate how technologies and elements interact.

Diesel-powered ships float on the materiality of the sea, but their engines need the atmosphere's oxygen and fire's spark to combust fuel and propel themselves across the ocean surface. These vessels move on the membrane between air and water, whereas drones and helicopters move within a single element while interrogating underwater lifeways below. This chapter has focused on one technology or another while isolating the way it conjoins with an element. In practice, however, the elemental-technology dyads are less discrete. Technologies intersect, interpenetrating multiple elements. Terms like *transtechnologies* and *transelementalities* more accurately incarnate the multiplicities that dominate human and nonhuman interactions.

Sea Shepherd's campaign ships are elemental technologies for "drawing out spatial distinctions between oppression and resistance, between centres and margins, between same and other, often refracted through the difference apparent between the land and the sea" (Hasty and Peters 2012, 665). With the campaign ship afloat, the sea is a tenuous, liminal element mediating the hazards of activism. Sailors, pirates, and enslaved people traverse the high seas and, in the process, their shipped bodies cross ethical, elemental, and legal boundaries (Ogborn 2000, 6). Like the drone and the helicopter, the ship's technical tendency is toward increasing speed, efficiency, force, and movement—all in the service of control. As this efficiency amplifies, the boundaries separating elements soften into a transelemental space of activist exteriorization.

Consider Sea Shepherd's small, fast, and agile RIBs, which offer key tactical advantages over their multiton vessels in the campaigns against whaling in Antarctic waters. A central tactic of these quick speedboats is prop-fouling, a monkeywrenching technique that involves intentionally entangling the propellers of whaling ships with polyurethane ropes braided with metal cables and steel junk (figure 3.5). To do so requires a choreography of elements and technologies. Once targets are spotted, volunteers don immersion suits, floating waterproof outfits that prolong

FIGURE 3.5. Sea Shepherd's rigid inflatable boat deploying a prop fouler in 2012. Notice the water cannon and the paintball splatters, a sign of another atmospheric technology, on the hull of the *Yushin Maru, 2*. Source: Sea Shepherd Conservation Society, Billy Danger / AP, 2012, reproduced in Chappell 2012.

their lives if they slip and cross the air-to-water membrane. The atmosphere and the ocean mediate while challenging their mission.

The RIB's journey begins with another transelemental technology, a crane that collects it from the larger ship, carries it through the air, and places it on the water. When the boat is held by the crane above the freezing waters, it is vulnerable to winds that can pick it up and flip it over, as happened in 2008, or bend and destroy the crane, giving the ICR whalers time to escape, as occurred in 2010 (*Whale Wars* 2010). When the boat is floating on the ocean surface, the drivers climb on ladders through the air, onto and out of the RIBs. Waves crest and rains descend as the boat hangs in the air. The pilots dangle on the ladder, the ocean and air meeting in this suspended weather-world (*Whale Wars* 2009b). This precarious deployment is situated between elements and technologies, an example of technics "ready-to-hand" (or *Zuhandenheit*; Heidegger [1927] 1962) yet also a "technics out of control" (Winner 1978)—a tenuous execution of blue governmentality to stop a ship and stay an extinction.

Prop-fouling without helicopter assistance is disorderly and prone to failure. Communication with RIBs is always difficult. Drivers wear water-resistant headphones and microphones while their hearing is obfuscated by bow-crashing waves, distracting sailing responsibilities, and surging tempests. In the trough of the swell, pilots only have two hundred meters of visibility. The helicopter adds miles of vision. In one instance in 2009, Sea Shepherd's RIB raced to the harpoon vessel to deploy a prop-fouler. The helicopter attempted to relay instructions, but the code phrase "Tora, Tora, Tora" was misunderstood, and they failed to emplace the weapon, instead throwing butyric acid bombs. They eventually heard the code and launched the prop-fouler, but it was too late—the opponent had escaped (*Whale Wars* 2009b). In this example, the RIB floated on the ocean's materiality while its buggy radio wave connected through the atmosphere to a helicopter materially lifted by the air's viscosity in an operation to salvage underwater cetaceans that were existentially bound to the medium of a thriving ocean. These transelemental mediations illustrate blue governmentality as an "athwart theory," a seafarer's take on practices and concepts that corrode and dissolve in water (Helmreich 2009a).

The ocean is more-than-wet, and the atmosphere is more than a little influenced by the seas. The ocean is a dissipative system, stable in its thermodynamic fluctuations. This dissipation as well as its diverse interactions with other elemental formations such as land, ice, mist, and wind define the ocean's "more-than-wet ontology" (Peters and Steinberg 2019, 304). Through its "elemental interdependence, interaction, and mutation," the "ocean exceeds its liquidity" (Peters and Steinberg 2019, 288). Mediation in these states mutates, transducing signals with sometimes ghastly consequences. In Sea Shepherd's case, the atmosphere is not the only element mediating the liberation of interior to exterior milieus. Admixtures of water and wind in the form of storm and spray, and a technology as simple as a crane, structure transelemental activism. This transelementality of ocean and air, and its interaction with technologies, constrain and permit any achievements of blue governmentality.

TRANSTECHNOLOGIES

Elements are in between. The boundaries separating elements, like the edges between technologies, bleed into each other. Digital information travels across these interwoven technologies. To understand a technology's mediation means to grok how it comports with the others in its network. Consider remote-sensing systems. They assemble along with satellites and a variety of airborne and water-based sensors to acquire information about illicit activities over vast and underregulated marine regions, from below the sea surface to near space. Autonomous underwater vehicles, floats, gliders, drifters, and ship-based measurements may contribute to Sea Shepherd's real-time campaign planning (Hafeez et al. 2018). "Drones," environmental scientists Hilde Toonen and Simon Bush write, "make the oceans legible" (2020, 131), but only when merged with other sensing, computational, and scribal technologies. Along with elemental thinking, these technological multiplicities afford new exteriorizations for activists and novel opportunities for blue governmentality.

One assemblage of technologies applied in Sea Shepherd's work is machine learning informed by drone images. In Operation Jeedara in 2018, Sea Shepherd flew drones in the Great Australian Bight, around Pearson Isles, to count and classify endangered Australian sea lions (*Neophoca cinerea*) and little penguins (*Eudyptula minor*). The 3D computer models they built with these drone images indicate population health and support arguments to protect the Bight from offshore oil drilling, like that proposed by the Norwegian company Equinor in 2020 (Taylor and Soliman Hunter 2020). Working with the Australian Department for Environment and Water and the University of Adelaide, the incorporation of recent technologies and new partnerships represents a less antagonistic and more collaborative approach between institutions and across technologies to achieve conservation objectives (Sea Shepherd Conservation Society 2018).

This was a relatively routine operation, but oceanographic fieldwork is often more speculative and precarious. Others, however, see the ocean as becoming knowable. Ocean cultural geographer Jessica Lehman argues that drones transform the ocean into a "frictionless field of data" (2017, 58), abstracting the sensed data in ways that prepare the ocean for governmental control, corporate capitalization, and industrial extractivism. I see the opposite in Sea Shepherd. Edges of extinction, illegality, and hand-to-hand combat on speedboats; the hazards of icebergs, freezing ocean swell, violent storms, and their transelemental synergies; and the absence of light and precarious radio wave connectivities load this ocean drone work with friction despite being patrolled by networked technologies.

Having to do more with less, activists have long incorporated diverse human and nonhuman actants into novel transtechnological assemblages. Some argue that paired with other technologies—3D modeling software, radar, thermal camera traps, and motion and seismic sensors—the drone "can greatly reduce poaching—but only in those areas where rangers on the ground are at the ready to use that data" (Snitch 2015).

Drones are but one important sensor system engaging a larger exterior milieu of technologies, collaborations, practices, and living beings.

Integrating drone praxis into transelemental, multispecies, and transtechnological networks, and combining this with a regulated networked public sphere and functioning legal system—as idealistically technoliberal as that may sound to some—is one example of a progressive interior and exterior milieu concretion toward biocentric justice. Effective conservation must concern itself not only with single or convergent transtechnological mediation but also loop back, in the fashion of Leroi-Gourhan's thinking, into the interior milieu first realized in the human hand, eye, and brain, and a broader interior milieu of tradition, law enforcement, and cultural values. What surfaces is blue governmentality—a never complete and elementally distorted mode of physical, tactical, and sensor- and code-based control over multispecies lifeways.

"PROMISING, BUT ALSO EXPERIMENTAL"

For Sea Shepherd, the drone—like their other elemental technologies—closes the gaps between interior and exterior milieus. Drone activism here emerges from a nature realism with a technical tendency toward improving situational awareness, proximity to endangered species, documentation of illegality, and the production of spectacle—all of which rely upon elemental mediations. The interior milieus of human conservationists, transelemental mediations, and transtechnological sensors influence the drone's technical tendencies toward atmoactivism. They come into exterior milieus of whaling, poaching, conservation, science, televisual entertainment, and public mobilization. The historical directionality of this tendency is not yet known. As a group of drone cetologists warn, the drone's technical tendencies are "promising, but also experimental" (Ferguson et al. 2018, 150–51).

Sea Shepherd has qualities that are enviable from a strategic perspective: committed volunteers, charismatic leadership, technological acumen, tactical wherewithal, and luck. These are factors that activist organizations might attempt to emulate. Sea Shepherd's volunteers are conservationists first. Their element-exploiting technical skills are secondary to the multispecies empathy that comes with the nature realism of their interior milieu. It is not only the thrill of moving quickly in the air, across the sea, or in forceful engagement with an enemy that motivates the volunteers, but it is necessary to achieve their conservation goals. Here, technological innovation and political aspiration are interlaced, each pulling the other in a technical tendency of conservation. Any conservation gains are made possible by the originary technicity of enhanced mobility, speed, and force through the elements.

The drone's relative affordability, agility, and ease of use have made it a revolutionary tool in oceanic science and activism. But the drone has yet to revolutionize conservation itself. Understanding this means grasping the complexities and contingencies of blue governmentality. Results of blue governmentality may be technological enfranchisement or circumscribed deliverance, with biofouling and wave action eroding these fragile formations. Blue governmentality offers a paradox of synergistic simultaneity, more a confusion of liberation and subjugation than an easy reconciliation. This blue governmentality, however, contradicts Sea Shepherd volunteers' ideal ocean devoid of ecologically disruptive humans. Sea Shepherd as well as the Ocean Alliance from chapter 2 may not like it, but blue governmentality—even in its fragmented form—may be necessary to temporarily achieve some of their objectives. For Sea Shepherd, nature is not over, but it could be. For them, precise interventions with technologies are necessary to prevent extinction.

Technologies are liberated from interior milieus through technical tendencies that form and in turn are formed by exterior milieus. This exteriorization of technics is epiphylogenetic, existing between situated haptics and uncanny otherness (Stiegler 1998). A technology liberated from human feet, hands, and eyes, equipped with sensors and memory, the drone might someday become an epiphylogenetic other. In this equally promising and concerning experimental future, drone liberation in conservation might mean less human agency and more dependence upon autonomous technologies. As Leroi-Gourhan originally warned over fifty years ago, we "must get used to being less clever than the artificial brain that we have produced, just as our teeth are less strong than a millstone and *our ability to fly negligible with that of a jetcraft*" (Leroi-Gourhan [1964] 1993, 295, italics added). If along with this humility comes a warped blue governmentality and the preservation of some ocean biodiversity, Sea Shepherd and Ocean Alliance will continue to welcome these flying artificial brains into our shared airspace, collective future, and battered shores.

4

GOVERNMENTALITY
FLYING TO THE LIMITS OF
THE LAW AGAINST SHARK
FIN POACHERS

SHARK FINNING IN TIMOR-LESTE

In early February 2017, Sea Shepherd received drone footage from a female diver from Timor-Leste. The anonymous pilot usually flew her drone to catch fleeting glimpses of blue whales as they migrated past Dili, the nation's capital. But in this video, she documented a Chinese fleet transferring or transshipping sharks to a vast refrigerated fishing vessel or "reefer" a mere 250 meters off the island country's north coast. This was the *Fu Yuan Yu* fleet, including the reefer the *Fu Yuan Yu Leng 999*. Chronic offenders of fishing laws, the crew of the *Fu Yuan Yu* are pirates, but instead of loot their booty is fish. The fleet is owned by Honglong Fisheries, an alter ego of Pingtan Marine Enterprises (Sea Shepherd Global 2018). Human trafficking, forced labor, debt bondage, forgery, bribery, and illegal fishing fill Pingtan's rap sheet (Marcus Aurelius Value 2017). Indonesia's Supreme Court found evidence of "torture ships" in their fleet (Barker 2018). For this reason, they were kicked out of Indonesia, following a raid of their offices by the Indonesian Armed

Forces. Their stock cratered. Pingtan's rebound was to harvest sharks from Timor-Leste and other small, biodiverse-rich countries like Ecuador without adequate patrol vessels. Despite their infamy, for a fee of US$312,450, the fisheries administration of Timor-Leste awarded the *Fu Yuan Yu* fleet the right to fish inside its sovereign waters for one year beginning in 2016.

We humans are eating our way through the sea (Probyn 2016). Ninety percent of fish stocks are fully exploited, according to the United Nations (Kituyi and Thomson 2018). No one is certain of the number, but with somewhere between 200,000 and 800,000 ships, the Chinese sail the largest fishing fleet in the world. Almost three thousand of these vessels are geared for long-distance journeys—a number three times larger than those of the next four countries (Japan, South Korea, Taiwan, and Spain) combined (Urbina 2020). The need to secure food for 1.4 billion people, the decline of local fisheries due to pollution and overfishing, the dietary desires of a growing middle class, and the projection of geopolitical power have encouraged the Chinese government to subsidize this fleet with billions of yuan. Government handouts to fishing capitalists are rife throughout the world. In a study of 152 ocean-faring countries, US$22 billion of state coffers is spent filling fishing holds. Such subsidies keep market forces at bay by covering economic costs. The result is that wasteful fishing remains afloat and sustainable fishing out of reach (Sumaila et al. 2019). Without state assistance, fuel costs to trawl the most distant seas would be astronomical, and such fishing would end. "It's not just the exotic stuff they're after," confessed Ian Storey, a security researcher at the ISEAS Yusof Ishak Institute. "It's pretty much anything that swims in the ocean" (Flitton 2017). China is the world leader in illegal, unreported, and unregulated fishing (Urbina 2020). For this reason, they are a primary concern for marine conservationists.

Open-water transshipping is illegal because it makes it difficult to regulate and track fishing. Consequently, soon after receiving the drone evidence from Sea Shepherd, the Policia Nacionale Timor-Leste (PNTL) and two officers from the Australian Fisheries Management Authority boarded the *Fu Yuan Yu* vessels and commenced an inspection that lasted three days. On board were as many as forty-three tons of sharks, many of endangered species. *Fu Yuan Yu* had authority from Timor-Leste to fish in these waters—but not in the way they were, nor for these fish. The government quickly surmised that they had contravened Timor-Leste law, violating regulations for transshipment, endangered species, and national fishing ordinances. Sea Shepherd and other watchers were certain that a conviction would be forthcoming for these notorious poachers. They wanted them off the water, their ships impounded, fines levied, and the fishers imprisoned or sent to China. But this was not to be. Further adventures awaited these fishing pirates, as Timor-Leste let the fleet sail with no suspension of their rights to fish in their sovereign waters. This decision was to be significant for another small maritime country and its fish.

The *Fu Yuan Yu* fleet continued shark fishing, sailing 17,000 kilometers east to Ecuador, to the outskirts of the Galápagos National Park, a sanctuary that protects the world's highest concentration of biodiversity. Fish biomass here is 17.5 tons per hectare, twice as much as the second highest area documented, the Cocos Island National Park in Costa Rica (Howard 2016). This maritime zone has the greatest abundance of sharks in the world, with an economic impact of about US$5.4 million per living shark—versus US$200 for a dead one—over its lifetime in tourism-related income (Howard 2016). At nearly 8,000 square kilometers, the park's immense size is a challenge to monitor. "Resources are limited," Ecuadorian marine ecologist Pelayo Salinas (2017) says. "The bad guys are every day making more money. Patrolling is expensive, especially for a county that is in an economic crisis." Salinas attempts to keep up with the poachers' technology through crowdsourced funding, asking anybody to help purchase "two kick-ass, commando-style chase boats to keep the bad guys away from the Galápagos Marine Reserve" (Salinas 2017). At this writing, the campaign had raised only US$22,130 of the needed US$200,000. During this time, the Chinese have had as many as 340 fishing vessels prowling the perimeter of the Galápagos National Park (Urbina 2020).

After leaving Timor-Leste with impunity and forty-three tons of frozen shark carcasses, the gargantuan reefer the *Fu Yuan Yu Leng 999* was identified inside the Galápagos National Park. They made a mistake: they left the vessel's automatic identification system (AIS) on, and it beamed its location to park authorities. "To us that seems like madness," said Jorge Duran Herrera, an Ecuadorian naval captain (Zimmer 2017). Reverse engineering the AIS data showed that the *Fu Yuan Yu Leng 999* sailed east into the park. On August 13, 2017, Ecuador seized the vessel and with it three hundred tons of fish, most of them sharks, including endangered hammerheads (*Sphyrna*) and threatened silky sharks (*Carcharhinus falciformis*). This tonnage equates to roughly 48,000 dorsal, pelvic, and pectoral shark fins to be traded in gray markets in Asia, where it is an expensive status-symbol food. Young and baby sharks were among the catch, leading some to think many of these sharks came from within the Galápagos National Park, a shark nursery (Zimmer 2017). Much of this shark tonnage was from Timor-Leste. The Ecuadorians were less lenient than the Timorese. For violating its general criminal code against transporting or possessing endangered species, Ecuador imprisoned twenty crew members on sentences of one to three years, the massive reefer ship was impounded, and they were given a US$6.1 million fine (Bonaccorso et al. 2021).

With astounding mendacity, the remainder of the fleet sailed west. Using AIS, Sea Shepherd tracked the *Fu Yuan Yu* fleet back into maritime Timor-Leste. On September 5, 2017, at the contested ocean border between Australia and Timor-Leste, Sea Shepherd documented *Fu Yuan Yu* ships soaking their nets and towing in their catch. Using drones, Sea Shepherd could see that shark bodies and coral were being pulled from nets and bludgeoned with wooden mallets.

Gary Stokes, captain of the Sea Shepherd vessel *Ocean Warrior* that was trailing the *Fu Yuan Yu* vessels, gathered this evidence and went ashore to meet with the PNTL. In a little bar, he showed them the drone footage, and they could see that they "were being taken for a ride here" (Stokes, interview with author, March 3, 2021). The PNTL immediately decided to work with Sea Shepherd and search the vessels for incriminating evidence of illegal fishing.

On September 9, 2017, with armed members of PNTL, Sea Shepherd began their operation with a drone flight. Stokes flew the drone about fifteen meters above the foredeck of *Fu Yuan Yu 9608*. Through the drone monitor, he focused his attention on a crew member in a gray hoodie who was pulling in nets. As the fisherman cracked the skulls of netted sharks, he looked at the drone and appeared to laugh. Stokes thought to himself, "If I see that guy, I'm just going to . . . I want to kill him. He kept looking over at us and smiling. . . . I thought he was just enjoying killing sharks. He wasn't really a flavor of the month for me" (Stokes interview 2021).

With a GoPro camera attached to his head, a police escort, and a drone flying overhead "so we could have eyes above, just in case anybody was going to come out with a gun," Stokes boarded the *Fu Yuan Yu* vessels. On deck, Stokes separated the unsavory man from the rest of the crew and interrogated him. As they spoke, Stokes discovered that he, like Stokes's wife, was Filipino. They chatted and quickly built trust. He confessed to

FIGURE 4.1. Filipino whistleblower communicating to drone operator, "please come," from foredeck of *Fu Yuan Yu 9608*. Source: Courtesy of Sea Shepherd.

Stokes that in the Philippines they would be arrested for the amount of coral they were destroying. Stokes was impressed, thinking, "Wow, this guy isn't the monster I thought he was. He was actually quite compassionate. He was actually an engineer." At the end of their discussion, the man asked Stokes if he "got his message." Stokes had to confess, he had not (interview 2021).

After searching all fifteen vessels, Stokes and crew went back to the *Ocean Warrior*. Two hours later, the videographer came upstairs and said, "Gary, you want to have a look at this." The man in the gray hoodie had identified their surveillance drone in the sky and, after ascertaining that no other crew members were watching him, he bent over, laid down his hammer, and, with his finger, spelled out on the deck in English, "Please come" (figure 4.1). The man was a whistleblower, communicating an intimate plea to the drone and its operator (Stokes interview 2021).

Just before the drone flew over the ship, they had landed a massive hammerhead, and he wanted Sea Shepherd to find the incriminating evidence of the killing of this protected fish. "Incredible. . . . talk about sending chills down your spine. . . . The whole campaign started because of a drone," Stokes confessed (interview 2021). Later, the whistleblower continued his undercover work, collecting images of the Chinese crew tossing hammerheads overboard and sending this evidence to Stokes (figure 4.2).

Once onboard, the PNTL told Stokes, "You find one shark and I'll arrest all of them." They found a multitude. "That was it," Stokes told me. "He got on the radio and then basically said, 'Right, everybody is detained'" (interview 2021). Ten to fifteen thousand shark corpses lay frozen below

FIGURE 4.2. Chinese officers disposing of evidence of illegal fishing for endangered and protected sharks. Image collected clandestinely by Filipino whistleblower. Source: Courtesy of Sea Shepherd.

deck (figure 4.3). They were targeting sharks, which they were not permitted to do. On command from the PNTL, they retreated to a protected bay, where the crew remained below deck. One night, anchors were pulled, and the ships traveled purposelessly around the bay. To stop this mayhem with a threat, Sea Shepherd turned their powerful water cannon on at the exact height that would destroy the fishing boat's communication equipment and illuminated its spray with their spotlight. Eventually, the fleet of fifteen was impounded and the crew arrested.

FIGURE 4.3. Gary Stokes of Sea Shepherd and an armed member of Policia Nacionale Timor-Leste, with leopard shark (*Triakis semifasciata*) in the hold of a *Fu Yuan Yu* vessel. Source: Courtesy of Sea Shepherd.

After a nine-month investigation, the boats, the crew, and even the shark bodies sailed home for China. A payment of US$100,000 from an anonymous source was sufficient to secure their return (Barker 2018). Stokes told me what then proceeded: "Well, that was the biggest kick in the bollocks for the Timorese police because they really worked hard. They were passionate. They wanted to do the right thing. They had them— they had the evidence in their hand. They had photos of it. I don't know how that judge could sensibly rule against, unless he was paid off. So, the fleet left" (interview 2021). Remarkably, Timor-Leste claimed that no endangered sharks were present on board. Stokes disagrees: "When we boarded [the boats] . . . we instructed the [Timorese police] and they went to the forward freezer, and they dug down deep and they actually found hammerhead sharks" (Barker 2018). Sea Shepherd says the deal "reeks" and should force us to question the influence China has over its smaller, poorer, resource-rich neighbors (Barker 2018).

After a fiasco that included arrests in Ecuador and Timor-Leste, Sea Shepherd implored China, *Fu Yuan Yu*'s homeport, to follow suit. In a document sent on March 23, 2018, Sea Shepherd requested that top-ranking Chinese officials inspect the fleet upon arrival. They argued that the *Fu*

Fishermen extracting catch from nets with piles of shark on deck

FIGURE 4.4. Drone image of fishermen extracting catch from nets with piles of sharks on deck. Source: Courtesy of Sea Shepherd.

Yuan Yu fleet violated the fisheries laws of Timor-Leste, China, and Australia as well as the Convention on International Trade in Endangered Species of Wild Fauna and Flora (CITES). Drone images were central to their formal accusation.

Leading the exhibits were four still frames—each progressively closer to the accused vessels—taken by the anonymous female diver's drone. One of these is a composite of three high vertical images collected during the initial investigation in Timor-Leste in February 2017. Illegible red writing overlies images of four fishing boats docked at a reefer, while two vessels depart, beside an abandoned pier. Another image captures eleven of the vessels, and the final two are close-ups of the reefer's deck in which a hammerhead shark, a CITES-protected species, is visible— and circled and named in red marker. These drone images document the fleet's reign of illegality, bringing together conservationists, fisherfolk, activists, whistleblowers, and frozen sharks in an entrapment of life and death (figure 4.4).

After his work with Sea Shepherd ended, Stokes started Oceans Asia with Dr. Teale Bondaroff in 2019. Bondaroff traveled to the Southern Ocean with Sea Shepherd to stop the whaling discussed in chapter 3. Drawing from this ethnographic and practice-led research, Bondaroff calls this drone work "direct enforcement"—a direct-action methodology that seeks to prevent, deny, and coerce compliance with international laws. Like direct action, direct enforcement challenges the state's monopoly on violence by executing its own surveillance and violence through technologies and against the technologies of extinction. By assuming a policing position, direct enforcement requires states to respond with their police forces—otherwise their legitimacy is undermined. In this manner, "by taking the law into their own hands, activists incite states to act," Bondaroff and his coauthor claim (Eilstrup-Sangiovanni and Bondaroff 2014, 352). Such activism temporarily slows offending activity, allowing evidence to be collected for criminal prosecution and the public to be informed. But publicity is not the point of direct enforcement. Sea Shepherd's primary goal is intervention, not bearing witness or generating sympathetic images. Theirs is not activism in the networked public sphere but rather activism in the elements, to the point of physicality, and in the courts.

Sea Shepherd founder Paul Watson defends direct enforcement, saying,

> To me, [witnessing] means nothing. You are witness to an atrocity. How is that stopping anything? How can that be called non-violence? You are just allowing the violence to continue by doing nothing. I don't believe in protest either. I don't like protesting. Protesting is a very submissive thing. It is like, "Please, please, please don't kill the whales." I mean, you don't walk down the street and see a woman being raped, pedophilia, a cat being stomped, a whale being killed, and do nothing. That is just cowardice. They can call it bearing witness all they want, but as far as I am concerned, they are just afraid to do anything. (Dolman 2011)

Bearing witness—and the audiovisual collection of mnemonic mementos that usually accompany such exercises of seeing from afar—is insufficient for Watson. He is not a scientist and has no patience for a legislative process that might begin with the publication of scientific research. As Watson regularly intones, "We do not wave banners, we intervene" (quoted in Heller 2007, 4, 124). Sea Shepherd invokes the United Nations' 1982 World Charter for Nature to justify their interventionism, but the charter does not empower nonstate actors in legal enforcement (Preston 2012). Regardless, Sea Shepherd enforces their deep naturalist principles via direct enforcement. To them, bold action makes right: "We get away with what we do for the same reasons they do, governments are reluctant to enforce laws," Watson claims. "That's why I've sunk ten whaling ships and destroyed tens of millions of dollars' worth of illegal fishing gear, and I'm not in jail," he brags (Roeschke 2009).

This ecocentric vigilante justice might be called "coercive conservation" (Khatchadourian 2007, 1), and at times it works. "Interventionist activism has been markedly more successful than protest activism in reducing the number of whales taken illegally," writes law professor Anthony Moffa (2012, 209). During Japan's 2012–13 whaling season, only 103 whales of a quota of 1,000 were killed (Nagtzaam 2014, 670). Sea Shepherd's successes in curbing whaling in the Southern Ocean, for example, results from compounding factors. One little-acknowledged factor is the adaptive exploitation of the atmosphere and watery elements. Their "direct action," "direct enforcement," "coercive conservation," or "interventionist activism" creatively engages the elements of the ocean and the atmosphere as mediators for mobility, suspension, and force. This vigilant elemental activism depends on the use and direction of water, and the movement that air and water allow. Its tactics vary from the subtle to the spectacular, the comic to the profound. Drones are tools that allow scientists and activists to exploit elemental affordances and attempt to mediate vitality itself. This blue governmentality tries to control the dynamics of marine zoēpolitics. With drone exosomatics, diverse oceanic and human cultures touch and intermingle. Yet amid the connective affordances of technicity and elementality, the human inventions of laws—both their enforcement and corruption—continue to influence the efficacy of blue governmentality.

Governmentality is a theory of how a population's reproductive capacities are monitored and controlled. For historian Michel Foucault (2003), this meant the governance of people, but sociologist Thomas Lemke finds in Foucault's "government of things" an account of "the interrelatedness and entanglements of men and things, the natural and the artificial, the physical and the moral" (Lemke 2015, 13). Several contemporary studies of ocean management bring together a focus on the sea's materiality and a theory of governmentality (Boucquey et al. 2019; Lehman 2016; O'Grady 2019). Some of this work combines Foucault's scholarship with that of anthropologist Elizabeth Povinelli (2016) and her concept of "geontopower," which, unlike Foucault's biopolitical focus on arbitrations of life and death, fixates on the distinction between the living and nonlife—minerals, deserts, atmospheres, or oceans. Awareness of the vitality of the nonliving, not just whether it is dead, invites new forms of control but also novel slippages in this governance. In this manner, it is not only the elements that furnish the spatial constraints and realms of possibilities for governance; police enforcement, international maritime laws, and corruption are also factors.

Geographers of oceanic geontopower usually investigate globally sensed and aggregated data sets that are accessed at computers maintained away from exposure to the elements. For example, geographer Jessica Lehman (2016) recognizes the potential and the failures of the vast Global Ocean Observing System (GOOS) and its efforts to informationalize, or as philosopher Bernard Stiegler (2018) would say, to "grammatize" the ocean's exceptional materiality. Another perspective on oceanic geontopower comes from ocean geographers Noëlle Boucquey and colleagues (2019), whose investigation exposes how the data portals of marine spatial planning (MSP) both harden and let slip ocean ontologies, affording a political potential of care beyond ocean measurement. Likewise, geographer Nathaniel O'Grady (2019) theorizes the "elemental excess" in emergency response during weather emergencies. These studies survey the care/control matrix of conservation governance in an elemental milieu. Like these seeing-and-planning systems, the drone offers a diffracted view into marine resource exploitation. Yet in contrast to the satellite's macro view, the drone's technicity provides a physical closeness between piloting conservationist, drone, and marine species. However, the high-resolution evidence that the drone produces through its caring proximity is but one component of blue governmentality and its wavering efficacy in marine conservation.

Drone conservation is fundamentally discrepant from GOOS, MSP, and other distributed and distanced systems of oceanic governance. The drone's exteriorizations remain proximal to the pilot's body, requiring constant mental attention during its short flight, flowing with and against the changing elements to bring humans and nonhumans into closer contact. Situated in the turbulent materiality of the ocean and the atmosphere, pilots and drones enact real-time slippages of the ideals of grammatization. Between the ocean and atmosphere, amid their soaking and saturating materiality, the drone and the pilot open potentials for novel encounters with elemental forces as well as aspirations for life's control. The governance of flourishing that follows is contingent upon the capacities of the drone's relative technicity and elementality as well as the vagaries of human law.

Blue governmentality is affirmatively biopolitical—it attempts to arbitrate life and death in the desperate calculus of marine existence. This labor represents a variety of "green governmentality" (Rutherford 2007), a concept of ecopower or the biopolitical governance over organisms, but with an accentuation on the role of technologies and tactics as well as legal and elemental exigencies in the oceanic and atmospheric context. Combined with oceanic elementality and its emphasis on optical technologies, the terrestriality of this governmentality shifts the visible light spectrum from green to blue, a material-ontological, legal-discursive, sensed-computed, and always partial attempt to control ocean life. Unlike other readings of biopolitics, blue governmentality is inherently multiple, inconclusive, and prone to failures and false positives. Blue governmentality pairs the exertion of biopolitics with the ontopower of elemental fluidity and ecosystem perturbation.

In blue governmentality, the azure sky and cyan ocean—and the technologies and waveform communications that flow through these elements—at once afford and constrain the control of biopower. Understanding blue governmentality means appreciating how the elements mediate movement, communication, and vitality itself. It requires comprehending how technologies emerge from the body, amplify political intentionality, work across the elements, and network with other technologies in forming cross-species intimacies. It approaches the legal limits of governance in international waters. Blue governmentality is geared toward multispecies flourishing. The case of Chinese shark fin poaching in Timor-Leste in 2017 exposes the elemental and legal limits of blue governmentality—the theory of care/control in ocean/cultures.

CONSERVATION AND ITS DISCONTENTED HUMANS

Many marine conservationists and ocean activists believe that monitoring the sea with sensing technologies such as small, affordable, and nonmilitary drones is beneficial for managing threats to endangered species (Hafeez et al. 2018; Millner 2020; Toonen and Bush 2020). Several academic geographers disagree, however, with this rather anodyne observation (Lunstrum 2014; Sandbrook 2015; Duffy et al. 2019). Conservation geographer Chris Sandbrook, for instance, claims that villagers who live in surrounding conservation areas are persuaded into "conspiracy theories, suspicions, and fantasies, particularly when [drones] are used" (2015, 647). Tanzanians might confuse the drone with a mythic "supernatural creature [that] swoops down and rapes them," or Nigerians might conceive the drone as a weapon for forced sterilization (647). These scholars encourage conservationists to move away from policing and surveillance, laws and their enforcement, and to focus instead on the semiotics of poaching, its effects on local people, and its contribution to militarization (Duffy et al. 2019, 67). In this calculation, extinction is not a criminal but a social problem. It is a crisis not of biology but of representation—both semantic and democratic. Long-term conservation requires support from local communities, and therefore sustainable conservation must include structural changes to address income inequality around conservation areas. Conservation technologies—helicopters, camera traps, drones, surveillance cameras, smart fences, rifles, and the like—are unnecessary, distracting, and harmful props in a symbolic struggle for fundraising attention (Massé 2018; Sandbrook, Luque-Lora, and Adams 2018). Conservation should not be mitigated with originary technicity—technologies of vision, tracking, and enforcement—but with economic welfare and educational justice. Conservation technologies abrogate these moral obligations. In this reasoning, the livelihoods of local humans must be addressed before those of endangered species.

These critics would argue that conservation drones are yet another technology of negative biopolitics attempting to artificially "maintain 'natural' wildlife populations" (Avron 2017, 364). Conservationists are embroiled in a "politics of hysteria" wherein violence against animals and rangers is "exaggerated" (Büscher 2016, 979). Decriminalizing poaching—justified because of the social harms of interdiction and because it is a mode of traditional hunting—is advocated (Büscher and Ramutsindela 2016). Poaching does not warrant high-tech attention because it is "non-threatening to the human society" (Simlai 2015, 39). These conservation geographers might agree with anthropologist Conrad Kottak, who states, "People must come first" (1999, 33). These detractors are unified in their skepticism that technology is a solution to conservation.

While it has been true that for too long technologies have been seen as a salve for many social problems, the inverse is also problematic. As design theorist Benjamin Bratton writes, "We are often criticized for 'technosolutionism' but there is also 'political solutionism' which emphasizes discourse and human institutions over the determining factors of technology" (2021, 148). Political solutionism is antagonistic to the ethos of conservation, according to conservationist Eileen Crist, who writes,

The literature challenging traditional conservation strategies as locking people out, and as locking away sources of human livelihood, rarely tackles either the broader distribution of poverty or its root social causes; rather, strictly protected areas are scapegoated, and wild nature, once again, is targeted to take the fall for the purported betterment of people, while subordination and exploitation of nature remain unchallenged. The prevailing mindset of humanity's entitlement to avail itself of the natural world without limitation is easily, if tacitly, invoked by arguments that demand that wilderness . . . offer up its "natural resources"—in the name of justice. (2015, 93)

The colonialism behind conservation needs to be addressed, and I understand how nature, as we think of it, is receding. At the same time, I also share a consternation with those such as public ethicist Clive Hamilton (2017), who questions why, at this critical moment in the collapse of ecosystems and the demise of biodiversity, it has become so fashionable to critique conservation. Ecocritic Greg Garrard wonders why so many academics ignore the "unsustainable discourses of modern consumerism, but [focus instead on] environmental rhetoric? Why have so many postcolonial critics attacked the conservation policies of environmental NGOs? Why are ecofeminists seemingly more concerned about the rhetoric of 'population control' than the effects of burgeoning human populations on biodiversity and global climate? . . . When, in short, did ecocriticism come to focus so much on criticizing environmentalism?" (2012, 510).

The marine activists described in this chapter advocate for conservation, not its deconstruction. With their high-tech, pseudo-paramilitary drone work, they defend animal flourishing. Despite the overlaps shared by ecological degradation and poverty, their work does not address the racism, sexism, or classism of environmental injustice. The class and race aspects of their work are not lost on these activists. As noted by media scholar Chris Robé (2015), Sea Shepherd's volunteers are predominantly young and white and express awareness of the racial qualities of their attacks on working-class Japanese whalers and Mexican poachers, for instance, as featured in chapter 3. This is a serious detail, but only one among many in the complex power dynamics on these high seas, where elemental dangers of swells, storms, and icebergs and elemental technologies of telecommunications, ships, and drones influence the military-esque maneuvers of this activism. For these actants, nonhuman conservation is an issue of thanatos and bios, not race and class. The immediacy of the extinction event is such, for these activists, that cleaving these dyads—like separating oceans and cultures—becomes paramount.

Throughout a decade of pursuing shark poachers, Sea Shepherd has been criticized for ignoring the rights of local fisherfolk. While following illegal shark finning operations in the Galápagos Islands in 2011, Watson responded directly to this critique, uttering with disdain,

> We are always criticized for going after poor fishermen. Well, these poor fishermen are wiping out sharks and turtles and everything. And so I find that people are always just being used in that way to justify this ongoing destruction of the environment and other species. And you know what I find really bizarre about it is nobody really cares about those fishermen. They don't have safety equipment—200 of them die every year. We don't think about those people, we don't care about those people, except when conservationists intervene and they say those conservationists are taking jobs away from those poor people. (Dolman 2011)

Although calls for economic justice for vulnerable human communities are frequent in conservation geography (Neimark 2019), opprobrium is rarely uttered with the same verve for the loss of wildlife. Within the justice framework, drone conservationists might argue that "we need to consider justice more broadly, as something that depends on simultaneously upholding the common good of the human and the non-human" (Kopnina 2016, 423). In this distributive justice, the risks and benefits of technoscientific progress are equally distributed across the animal kingdom (Rawls 1971). Both the benefits of drone systems (remote sensing, personalized mapping, creative interrogation, etc.) and the risks of the drone industry (mining of cobalt and rare earth metals for transistors, fracking of methane for electrical power, releasing noxious gases in their recycling, strip mining for copper wiring, etc.) should be equally distributed in both human and nonhuman worlds. The life-preserving potential of drones would positively impact nonhuman thriving, and the negative consequences of producing drones would be felt in human worlds. We might call this principle *multispecies justice*, or the repositioning of justice to encompass all beings (Tschakert et al. 2020, 2). Practicing conservationists could support multispecies justice with their drone piloting and mission execution. For ocean activists, drones are exosomaticizations of interior milieus that can be deployed to sustain life.

Many scientists agree with Sea Shepherd's nature realism that there is little time to waste in processes of political debate. Biologists Gerardo Ceballos, Paul Ehrlich, and Peter Raven (2020, 1) argue that extinction reduces Earth's ability to provide "ecosystem services" to humans so the "ongoing sixth mass extinction may be the most serious environmental threat to the persistence of civilization, because it is irreversible." Genetically valuable individuals within endangered species are reproducing less and dying at an alarming rate. Habitats are disappearing. Overfishing, pollution, and climate chaos are increasing. "Novel ecologies" will not offer the biomass, resiliency, interspecies dynamics, and nutrient densities that natural selection, growing populations, and life requires (Steffen et al. 2015; Valentine et al. 2020). The solace found in cherishing *any* vitality that might persevere on a "damaged planet" will be short lived (Tsing et al. 2017).

From the perspective of nature realists, any effective short-term efforts, including those that are armed and highly technologized, are necessary. All means—technological, atmospheric, militaristic, and directly confrontational—are on the table at this critical junction of the sixth extinction and the fourth industrial revolution (Braidotti 2019a). Parks, policing, drone surveillance, and other "political technologies" of conservation (Lorimer 2015, 191) are praxes that balance care with control. With ever-shifting elemental, animal, and political influences, open-water environments are notoriously difficult to care for—and even more challenging to control. Well-intended and half-successful attempts with drones are made to preserve flourishing, distinguishing life from death.

Sea Shepherd's ideal is that their activism will "lead to more just and ecologically sustainable practices of knowing—to a mode of governing with care" (Braverman and Johnson 2020, 20). The simultaneous presence of both care and control may not be the contradiction one might assume. Control and care are intertwined. Enhanced technologies provide opportunities for new types of care that ratchet up control. Laws and science too are interlinked: regulations limit science, while science at times informs laws. Governance starts with ordinary technicity, is moderated by elementality, and is made present by enforcement. These simplistic diagrams map the actants, configurations, and complications of blue governmentality.

Blue governmentality is a form of multispecies care and control that is circumscribed by elemental turbulence, technical maladaptations, and human self-interest. Fueled by existential dread, comforted by technological idealism, and emboldened by deep naturalist fundamentalism, Sea Shepherd's blue governmentality represents a leading edge of technical tendencies aimed at an actively receding horizon of control through enhanced speed, force, and movement. In terms of originary technicity, the drone offers an extension of a human phenotype, suspending sight and airing out motility. With this capacity, Stokes flew above the *Fu Yuan Yu* fleet and depicted the slaughter and transshipment of sharks. The drone precipitated engagements between the PNTL and the pirate fleet (twice), as well as between a Sea Shepherd activist and a Filipino whistleblower. Elementally, the air provided the drone's lift and transparent skies for its documentation. The ocean furnished a smooth vector for the transportation of activists, industrial fishing vessels, and the transition from living to dead sharks. However, the care of marine conservation expressed by the PNTL, the courts of Ecuador, and Sea Shepherd was not matched with the Timor-Leste government's juridical control of the Chinese-owned fishing fleet. A bribe sprang the pirates from the foreign port.

The limitations on conservation are many: species competition, niche collapse, generalized entropy, climate change and human growth, disinterest, incompetence, and underfunding, to name several. The applications of conservation technologies are few: tracking, depicting, and physically intervening. The drone's usefulness depends on human systems subject to political and public persuasions. Conservation geographers who are skeptical of technologies also understand these limitations. For example, geographer Francis Massé (2018, 60) regularly encountered drones in his ethnographic work with conservationists at the Mozambique and South Africa borderlands. These drones, however, were not deployed to stop poaching. They were photogenic props, along with images of slaughtered wildlife and vulnerable rangers, in online fundraising campaigns. If successful, funds will be raised and spent on more drones that might militarize conservation, terrify local people, and distract from the social work of conservation. A loop of human harm and techno-militarization will proceed—or so goes the logic of these critical conservation geographers (Lunstrum 2014; Sandbrook 2015; Massé 2018; Sandbrook, Luque-Lora, and Adams 2018; Duffy et al. 2019).

They depict a dark, dominator governmentality—not the "left governmentality" Foucault (1977) eventually supported post–*Discipline and Punish*. While critics of governmentality in the human domain often document *subjection* (Butler 1997; Kelly 2009), the stabilization of subjectivity for purposes of control, blue governmentality aligns more with Foucault's later conceptualization of governmentality as *subjectivation*, or an affirmative liberation of subjectivity from the domination of previous state, religious, and monarchical systems (Foucault 2003; Fish and Follis 2016; Dean and Zamora 2021). This affirmative governmentality creates more than the space for personal performativity among repression; it is also about life and death. Bratton writes, "Control does not equal oppression; read your Foucault better. Control is also . . . the freedom not to die early and pointlessly" (2021, 146). Enforced by technologies, managed by regulations, limited by the contingencies of the elements and technologies, this is the flourishing by control of blue governmentality. Despite its faults, blue governmentality is for animal self-determination. It seeks to use technologies and management to protect the autonomy of nonhuman life. The conservation drone of blue governmentality is one that is less an engine of spectacle and means of securitization, and more an imperfect tool for defending the subjective thriving of marine species.

Cultural geographers critical of conservation technologies have kinship with political ecologists and science and technology scholars who claim that there is a slippery slope between scientific quantification and the commodification of nature. In one example, they argue that while whaling has ceased in some contexts, the rise of whale ecotourism is yet a continuation of the "commoditized whale" (Turnhout et al. 2013, 157). Whale watching does "not in fact protect them from markets but brings them to new markets and exposes them once again to human disturbance" (157). The "disturbance" of a fatal blow from an explosive-tipped harpoon is distinct in kind and degree from the chase, harassment, gazing, and Instagramming by tourists from a nearby vessel. Both are disruptive, but one is preferred by whales. The commodification of pictures of the whale, instead of its butchered parts, leaves the whale alive, not dead. Academic arguments such as these fail a basic tenet that vitality is preferable to death. It is a politics of grievance against power over gratitude for being alive. Living with animals, and animals living with us, will require adaptation and resilience to our less fatal methods of coexistence. This might include the disruption of monitoring technologies such as drones and the incessant chatter of tourists and the whirl of their diesel engines. Such uncompromisingly cynical conceptualizations of control weaken critical realist progress toward flourishing together.

In this technophobic, negative governmentality, the simple, pragmatic question of how conservation technologies might help nonhumans is rarely posed. Instead, attention is placed on minimizing suffering to humans, and, as Crist wrote, the ocean is hailed to "offer up its 'natural resources'—in the name of justice" (2015, 93). This focus on drone discourse instead of drone technicity shifts the discussion from material efficacy to representative justice. Drone discourse is a powerful agent; it influences practices, generating the language needed to describe and enact power. Discourse can inspire the investments that finance development. In drone business, imaginaries of industrial revolution and teleological inevitability dominate (Jackman 2016). But materiality and practices constitute formations that are important to distinguish analytically from discourse. A material approach to the drone frames it as not merely constructed through textual descriptions, oral seductions, mediatic hyperbole, and public performances, but also as a physical object that is lifted and contested by physics, technological affordances, technical skills, and political ambitions. It is a tool for intervention, emerging out of the evolution of originary technicity.

Following Sea Shepherd's lead in privileging action over representation, this chapter offers a material/practice-based approach to drone conservation. Sea Shepherd strives for physical metrics—lives saved, arrests made, enemies disturbed, boats sunk. When possible and helpful for their cause, they labor closely with armed police. With ships that look and function like state navy vessels, they bring to their operations the latest in tracking, navigation, research, and communication. Sea Shepherd, however, is not without its performativity and representational politics. A second-order effect of this confrontational conservation is compelling media. For raising awareness and funds, they are experts at generating film, television, memes, and social media of inspiring and courageous activists fighting the destruction of marine species. As we will read about in more detail in chapter 7 in the context of shark conservation by charismatic activists, here we witness the personalized politics of subjectivity or the positioning, framing, and embodied performance for a political cause (Reestorff 2014). This liberalization of the self is made possible by governmentality, the less dominating and more caring/control approach to biopolitics that Foucault later championed (Dean and Zamora 2021). Externalized through political bodies, the drone is both a media-producing camera and a character in these missives—a symbol of technological proficiency and a tool for serious commitment. The drone and its operators are at once producers and performers of technological competence, elemental situatedness, and multispecies empathy. A more accurate anthropology emerges when discursive and material studies are complementary.

On the hard Earth, poverty, hunger, unemployment, displacement, and environmental injustice motivate poachers. But the argument that poaching originates from social inequality must also consider elementality. That is, its economic logic might be particular to dry land. Putting up the initial costs for illegal and unregulated fishing requires significant capital investments. Purchasing a massive vessel and expensive tackle and communication equipment and hiring crew is not an option for poor, landlocked folk. Few human cultures have traditionally had access to deep ocean waters to exploit Antarctic fin whales, for instance, or South Pacific pelagic sharks. There is no known traditional cultural practice of vaquita fishing—no habit of accumulating forty-two tons of sharks in the freezing belly of a ship roaring with diesel power. Impoverished people often accept underpaid, inhumane, and sometimes enslaved labor aboard fishing vessels, certainly, but marine poaching is an act of capital and captains, not the crew of desperate people. Industrial fishing on this scale is not an Indigenous culture to fight for and protect. Our theory building needs to attend to the technological, elemental, and economic materialities of our research participants. Blue governmentality is a proposal to focus on how power moves through the air, over the ocean, and into and out of courts, generating the economic, technical, and existential capacities of particular ocean/cultures.

On land, commercial poachers are organized as distributed networks and utilize firearms, GPS, and mobile phones. On sea, much more sophisticated technological and social systems are required, including ships, communication technologies, and costly bribes. Sea Shepherd's vigilante justice, paramilitary stance, and appropriation of surveillance technology is in keeping with the desperate pursuit of ever-decreasing fish stocks. As chapter 3 explored, Sea Shepherd's drone was shot out of the sky over the Sea of Cortez, their ships rammed and sunk in the Southern Ocean. Terrestrial poachers now use helicopters and drones. Drone conservationist Lian Pin Koh (n.d.) discusses this technological evolution: "poachers could use this technology to locate threatened wildlife. I expect criminals will eventually be using drones for poaching and hunting." As a practitioner, Koh and his collaborators provide experiments, technological prototypes, and practical protocols for how to bifurcate the drone future toward life, not death. Sea Shepherd's conservation drone is an example of one such drone geared toward prolonging bios. What results is a compromised blue governmentality, where technologies, the elements, and courts provide platforms for truncated forms of care and improbable forms of control.

This governmentality will not be easy, entirely effective, or without conflict. Koh argues that drone conservationists need to keep up with this arms race or they will be surpassed. Contrary to the belief of cultural geographers who are critical of conservation technologies, this is not a battle between an elitist Western desire to protect charismatic mega-fauna and poor victims living around conservation parks, but rather a fight against organized poachers, as primatologist Jane Goodall (2015) argues. Advocacy for the Indian one-horned rhino in Kaziranga National Park in Nepal succeeded because of its hard-line approach (Balmford 2012; Saikia 2009). Convincing evidence shows that strict protection of wildlife, with limited access for ecological tourism, is effective for the protection of biodiversity (Kopnina 2016, 423). Regular monitoring is essential to ensure wildlife protection, particularly for threatened species (Kays, McShea, and Wikelski 2020; Linchant et al. 2015). Drone practitioners and theorists Margarita Mulero-Pázmány and colleagues (2014) contrast attempts to limit poaching—horn control, laws, education, rural development, demand—with what they see as the potential benefits of drones as tools of surveillance, support, and deterrence. Geographers Audrey Verma, René van der Wal, and Anke Fischer (2016, 85) argue that technologies aid conservation by advancing an aggressive approach and by enabling conservation managers to control their narrative and representation. The World Wildlife Fund developed an algorithm that predicts with 95 percent accuracy the location of poachers and threatened animals. Drone data trained this system with pictures of animals and people in the bush (Andrews 2014). These are terrestrial examples, but ones that disclose the potential of technology-assisted, terrestrial, green governmentality. Blue governmentality, despite its limitations, can be one vital component for widening the reach, visibility, and force of conservation through the atmosphere, across the oceans, and into courtrooms.

For critics, technologies like drones have totalizing abilities—earth-transcending mobility, complete autonomy, and artificial intelligence capable of identifying and controlling people and other animals. This book's examples of how drones are elementally configured and politically muted show how they have few of these overarching capacities. Far from the "god-trick" of seeing everything from nowhere (Haraway 1988), as is often claimed (Feigenbaum 2015), the drone's mobility and vision are not dominating but dictated by situated praxis that is prone to elemental perturbations and technological inefficiencies that lead to failure, distortions, and mission compromise. The elements lift but also crash drones; the electromagnetic spectrum carries but also glitches communication; and the ocean medium suffocates as well as supports life. Drones capture high-resolution images that are not always given weighty consideration by judges. In these matters, I side with scholars who identify drone work not as a terrifying surveillance but as an embodied practice that is personal, that is community oriented, and that harbors the possibility of protecting vivacity (Helmreich 2016; Hildebrand 2019; Kaplan 2017; Millner 2020; Peckham and Sinha 2019; Radjawali, Pye, and Flitner 2017; Vertesi 2012). Ocean drones forge data intimacies within conservation milieus that have the opportunity—but not the certainty—to broaden care across the species divide. With this embodied and in sight, we can be more cautious in our technophobia and bolder in our conservation praxes.

WET WILDERNESS

This book covers a rich range of drone applications, some of which prioritize minimizing the harm associated with drones, such as crashing them into rookeries, as you may read in chapter 6, others toward maximizing the ecological benefit of drones, such as catching illegal fishers (Wang, Christen, and Hunt 2021). *Oceaning*'s cases of drone science and activism span from the highly regulated research of a US federal agency in chapter 5 to the clandestine labor of activists pushing the technological limits of drones in pursuit of whalers and fishers in chapters 3 and 4. Drone technicity is limited by elementality, pilot competency, and technological sophistication. While "geofencing" software embedded in some drones by the manufacturer can stop a drone from going near or over certain areas, regulations on where one can and cannot fly limit only those pilots who elect to lawfully fly. Value judgments are involved in all piloting operations. How far away, high, or where a drone is—the qualities regulated by most nations' drone laws—can be interpreted divergently by different pilots. Furthermore, the drone activists whose nature realism may make them feel less bound by regulations may consider the existential stakes of their piloting broadly, not only in terms of the safety of who might be injured by a wayward drone but also for those animals whose flourishing might be protected by a quick, rather dangerous, and probably illegal flight. It may take a crook to catch a crook. Considering this, the affirmative approach taken in *Oceaning*, and because of my argument for a relative technicity toward speed, mobility, and power from afar, I am in no position to make a value judgment about where, when, or why a drone should fly—or whether the drone regulations are just and equitable. For many of the book's subjects, both human and nonhuman, this is considered an issue of life and death—a concern that may transcend prudence or jurisprudence.

The ocean is a wet wilderness. Distant from onshore governments, the sea seems unculturable. The ocean and the atmosphere, their vast size and ownership indeterminacies impede governance. Blue governance and its surveillance technologies are required, its enactors believe, to apprehend violators of conservation regulations and reverse oceanic decline. This is performed not only for streaming television and social media audiences, but in clandestine operations in the dead of night and the middle of the ocean. In this work, the deep-ecological ideal of an ocean free of human and technological disturbance is enacted through the supervision of shark fishing, for example, with less-than-ideal sensors, difficult-to-enforce laws, and across the perturbations of air and ocean. These activists and their collaborating police forces struggle to overcome the limits of their technologies, the weakness of international laws, and the challenges posed by the elements. These are the potentialities and tests of drone conservation.

Fishing in international limbos between nations, transshipping and therefore hiding their catch in reefer vessels, and skirting laws regarding bycatch, the *Fu Yuan Yu* fleet tried to escape regulatory oversight. Their flagrant escapade almost failed. Confrontational conservationists and police collaborators gathered evidence via a drone and a forensic search and seizure. Ecuador elected to fine and arrest crew members and captains, while Timor-Leste awarded a fishing license to a documented criminal and twice let them go with tons of sharks in their holds. Drone mobilities provided by originary technicity and elementality came together to formulate compelling evidence, but resulted in a miscarriage of blue governmentality. The drone's technicity allows for the collection of incriminating evidence; what the courts do with that evidence is another issue.

This chapter has introduced enforcement, punishment, and corruption as aspects of blue governmentality and its subversion. Conservation's objective is to save other species, not humans. It does so by aligning relative technicity and elementality toward partial governmentalities. In the process, technicity, elementality, and governmentality each bring into contact nature and culture, or in this case drones, ocean worlds, and activist aspirations—or what can be considered the separate but complementary aspects of ocean/cultures. The consequences of ocean/cultures would be an increasing reliance upon conservation technologies for marine thriving. But this is not happening: neither technology nor the law saves sharks from future netting and de-finning. No deterring punishment was given in Timor-Leste. Despite compelling evidence produced by a drone, the *Fu Yuan Yu* pirates sailed for home. The elementally and legally contingent efficacy of drone conservation clashes with the drone depicted in critical conservation literature. There we witness a drone with absolute power to militarize conservation and representation itself. A humbler drone is the more honest drone.

5

SAILDRONES AND INSTRUMENTED SEALS

It is the summer of 2017, and scientists from the US National Oceanic and Atmospheric Administration (NOAA) land at a rookery on the northeast side of St. Paul Island, the largest of the four Pribilof Islands of Alaska, between the United States and Russia. The scientists are here to capture sixteen mother northern fur seals (*Callorhinus ursinus*) and their pups, take blood samples and weight measurements, and instrument them with dive and geographic trackers and video cameras—sensors that will collect data about where, when, and how the seals hunt.

A Saildrone—a seven-meter-long solar-powered and wind-powered ocean surface drone with a rigid five-meter-high sail—follows the seals as they forage. Headquartered on San Francisco Bay, the for-profit company Saildrone Inc. is ambitious. They envision a fleet of a thousand rugged surface-sailing drones saturating the ocean. A pivot from the successful pursuit of the land-sailing speed record, Saildrones can get to any destination in the world within thirty days from the closest shore. If it avoids biofouling (marine growth on the hull) it can travel indefinitely, perpetually relaying information to terrestrial servers, staff, and the scientists or firms who employ them. Their objective is to "become a vital enabler of a near-real-time planetary scale monitoring system we call 'The Quantified Planet.' . . . After all—we can't fix what we can't measure, and we can't prepare for what we don't know" (Saildrone n.d.). In this operation, the Saildrone echolocates schools of Alaskan or walleye pollock (*Gadus chalcogrammus*), the seal's most important food (Lowry, Frost, and Loughlin 1988) and a species that is also highly desired by humans—it is considered one of the most valuable fish in the world, a favorite for McDonald's, Arby's, Long John Silver's, and other fast-food outlets' menus (Bailey et al. 2000).

There may once have been as many as twenty billion walleye pollock in the Bering Sea (Bailey 2011, 2). Despite this whopping population, pollock stocks can crash. Technologies such as stronger polyester nets, faster diesel engines, refrigeration, and echolocation have made fishing more effective and deadly (Urbina 2019). Lacking these technologies, seals lose to humans in the competition over pollock. Although hunting seals for their pelts originally decimated their numbers, overfishing for pollock now most affects the livelihood of seals. Today, the seals remain vulnerable, with only 650,000 individuals remaining, a third of what their population was in 1980 (Gelatt and Gentry 2018). Complicating matters in unknown ways is climate change. Bering Sea ice is melting and moving north, closer to the North Pole, and with it the pollock.

It is around but not directly into this problem that NOAA sails themselves and drones to the Pribilof Islands. Their research objective is to discover what the seals eat, how often, and how much energy is required to find and dive to catch pollock. The researchers return in a week, repeat the capture, and collect the sensors and data about the seals' and pups' health. In a warming Bering Sea, this project entangles researchers, seals, and drones in an intimate relationship, the reciprocities of which should be questioned.

Capturing a mother seal is difficult. Males defend their harems. From a protective and camouflaging capture box, the scientists study the personalities of the males, identifying which ones might allow them to get close to the females (figure 5.1). While they wait, pregnant mothers give birth on the gray and green stones of the shoreline. Dr. Carey Kuhn, the leader of this project, described to me the process of grabbing the seals: "There'll be a point when the animals are aware that we're in close prox-

FIGURE 5.1. Dr. Carey Kuhn scouting for a female seal to net from a capture box. Source: NOAA Fisheries.

imity and then they'll start to get nervous and start to move to the water. And so, we have to find that fine line between how close can we get versus not getting so close that we scare everybody. So, once we're ready, we'll jump up and we'll run and chase that specific individual and put her in a net as quickly as possible" (Kuhn, personal communication with author, March 27, 2020). No sophisticated technologies mediate this intimacy.

Historically explored by science and technology scholar Etienne Benson (2010) and examined in the contemporary period by sociologists Jennifer Gabrys and Helen Pritchard (2018), sensing is spreading from human-embodied to animal-instrumented practices. Science scholar Natalie Forssman (2017, 18) reads networked seals as "outsourced laborers" in scientific work within austerity science. Capturing and fitting seals with sensors is more affordable than hiring large research vessels and a human crew. Instrumented and freed to forage the ocean depths, but destined to repeat the dangerous procedure when they are recaptured to collect the sensors, the seal becomes a "knowledge delegate" or a "bodily worker" (Forssman 2017, 189) in a scientific experiment with little evidence that their labor assists in their survival. As Forssman observed, "seal scientists spend their time sneaking and crawling through the shit, seaweed, and sand of the shore, trying to gain purchase on their research subjects" (2017, 26–27). These scientists are extremely careful not to injure the seals. However, Forssman compares the physical effort by scientists to tag elephant seals (*Mirounga*) to the work of seal hunting. Both hunters and scientists "exploit the onshore vulnerabilities of these bodies and seek to capture and characterize them as repositories of energy stores" (Forssman 2017, 105). While the human scientists receive data with which to manufacture their reports, what the seals acquire is more obscure.

After attaching the trackers, Kuhn immediately logs in and monitors the seals' whereabouts: "I am always looking to see where they are. As soon as tags go on them the tags are transmitting to the Argo satellite system. So, then I can log in on any computer and get their most recent locations" (Kuhn personal communication 2020). From handling the living mothers to overseeing their pursuit of prey, contrastive types and scales of interspecies intimacy emerge from these numerous methodologies, technologies, and networks.

Grafting sensors onto ocean species began in the 1960s, when the US military and its subcontractors began to develop sensing technologies for cetaceans—whales, porpoises, and dolphins. This "military-cetological complex," as Benson (2011, 75) called it, had its problems. Signals from tagged dolphins would vanish, battery power would end, antennas would break off, electronics would glitch, researchers would look in the wrong location, and wired animals would die or be eaten (Benson 2011, 76). Despite this, Benson argues that by "rendering individual animals locatable and identifiable, it enabled [conservation biologists and] park administrators to assert a fine-grained disciplinary power in the name of the preservation of wildness" (2013, 178). In actual practice, however, animal tagging—radio systems strapped on orcas, in Benson's (2010) example— was messy and contingent with place-specific variables that made the disciplining less than successful. Regardless of its efficacy, these conservationists carried the "very American idea" that their care would come to fruition through these technologies of surveillance and control (Benson in Hamblin 2013, 17).

Since these early troubles, the sensors' systems, batteries, and applications have improved, and many animals have been instrumented. Penguins, seals, turtles, tunas, albatrosses, all sorts of whales, and other seafaring species have been fitted with sensors, becoming, whether they like it or not, "platforms," "data collectors," "oceanographers," and "profilers" of the oceans and their worlds (Boehlert et al. 2001; Charrassin et al. 2002; Fedak 2004). Instrumented marine species can be used for extractivism as well as conservation. For example, penguins were fitted with GPS trackers with the hopes that their migrations and predation geographies would inform oil exploration about where penguins migrated and therefore where not to prospect around the Falkland Islands. These penguin "data gaps" opened the door for oil development (Blair 2022). So both negative and positive biopolitics are possible for the participating instrumented animals, birds, and fishes.

Back on St. Paul Island, the mother seals are also fixed with Crittercams, or video cameras that biologists attach via suction cups, glue, or other harnesses to the bodies of fish and underwater animals. With both entertainment and scientific objectives, they were invented by marine biologist Greg Marshall in 1986 and first deployed by the National Geographic Channel in 1996. By 1998, Marshall and National Geographic had deployed 169 Crittercams on twenty-two species of turtles, whales, and sharks—basically anything large enough to carry the kit (Marshall 1998). Crittercams provide data about habitat use, diving behavior, foraging, sociality, territorial and reproductive practices, and locomotion (Marshall 1998, 13–14). Marshall claimed that, for human viewers of Crittercam footage, the tool increased intimacy and empathy with the networked animals.

Marshall's original idea for attaching Crittercams came from watching remoras (*Remora*), the parasitic fish that cinches to the bodies of sharks and manta rays. His plan was to enable human vision to mimic the hitchhiking of remoras. The mimicry of an animal that connects itself precariously to sea life offers a promising avenue for science and technology scholar Donna Haraway. For her, in this "becoming remora . . . we have left the garden of self-identity and risked the embodied longings and points of view of surrogates, substitutes, and sidekicks" (2008, 253). For Haraway, Crittercams give marine mammals and fish the opportunity to self-represent in a visual idiom comprehensible by humans. A clumsy yet coordinated dual embodiment proceeds, wherein scientific vision and animal bodies are conditioned by each other. An alien closeness occurs. A drone adds another component to this more-than-human kludge.

With the seal self-revealing its predatory map and Crittercam and dive trackers attached to its body, the Saildrone began to track the instrumented seals. Like Crittercam, the Saildrone enhances and obfuscates

FIGURE 5.2. Automatically collected "dronie" of Saildrone and harbor seal.
Source: NOAA.

relationships between humans and nonhumans. On its journey through the Bering Sea, the Saildrone encountered others. A rorqual whale (*Balaenopteridae*) was observed and automatically photographed. A harbor seal (*Phoca vitulina*) took a break on the Saildrone's stern (figure 5.2). The Saildrone moves by wind and measures wind speed. It calculates the location and consistencies of ice, while also avoiding that ice. The sun provides solar energy for the Saildrone's sensors, and its sunlight also grows phytoplankton on its hull, which might affect its mobility. The Saildrone is elemental: it documents elementality while it is elementally sustained and suspended.

The Saildrone recorded that the seals foraged farther from the island than previously assumed, and the Saildrone consequently adapted, modifying its monitoring of the volumetric ocean. Contrast the Saildrone's proximal efficacies with those of a research vessel such as NOAA's 208-foot, 2,400-ton stern trawler the *Oscar Dyson*, and its advantages become clear. A Saildrone costs US$2,500 a day to operate. This includes the staff that guide, program, and track and process the terabytes of data it produces. Sailing a massive NOAA vessel can cost as much as US$50,000–100,000 every day for fuel and consumables. The Saildrone can get closer to the seals, silently stalk them, and produce little disruptive wake (figure 5.3). "Fish are presumably more likely to react to large ships than small, quiet USVs," or unpersonned surface vehicles, write members of

FIGURE 5.3. Saildrone in the Bering Sea contrasted with the 208-foot, 2,400-ton NOAA research ship *Oscar Dyson* in the background. Source: NOAA Fisheries.

the NOAA team (De Robertis et al. 2019, 2). The Saildrone is more responsive in terms of movement and rapid mission modifications and can be deployed for longer expeditions (Kuhn et al. 2020, 5). As a sensor platform, sailing athwart the elements of ocean, sunlight, and gusting air, the Saildrone's contribution is intimate yet incremental; it augments rather than revolutionizes possible proximities. Its efficacy requires the primordial and corporeal work of capturing seal mothers onshore.

Despite the Saildrone's advantages, theorists remain skeptical of drone oceanography. As explored in previous chapters, marine geographer Jessica Lehman (2017, 58) argues that sea drones transform the ocean into a "frictionless field of data," a process that makes scientific work less embodied. Other scholars argue that sensing technologies transform the unruly ocean into manageable space, rendering marine species into a data stream that can be controlled (Boucquey et al. 2019). For them, the volumetric matrix of the sea is knowable—or it is in the process of becoming so with advancements in autonomous vehicles, artificial intelligence, instrumented animals, smart buoys, and satellite and multispectral optics. This datafication will domesticate the ocean, affording techniques of material manipulation for marine bureaucrats. Applied oceanographers know better than to subscribe to these overreaching conceptualizations of human capacity. They know that smart buoys and other ocean sensors fail in these totalizing endeavors. Accord-

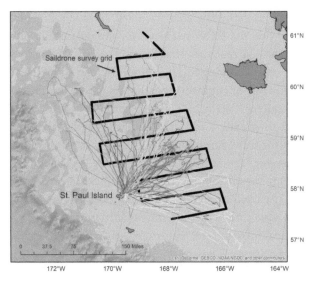

FIGURE 5.4.
Saildrone sur-
vey grid in black.
Foraging paths
of northern fur
seals in shades of
gray. Source: Sail-
drone 2019.

ing to marine anthropologist Stefan Helmreich, by "constantly trying to overcome and survive the sea, trying to deliver clean, mathematical lines . . . [ocean sensors] miss sea materiality" (2019, 19). The Saildrone conducts transects of oceanic space, overlapping with the seal's intuitive pursuit of prey (figure 5.4). These "focal follows" are a common methodology in movement ecology. Humans on foot, vehicle, or sea vessel track and record the paths of animals to better understand their home territory. But being autonomous and solar powered, Saildrones have fewer limitations in terms of range and duration than humans on foot or in jeeps. They are fast, highly networked, and can gather data sets humans have difficulty recording, such as acoustic measurements. Not needing fuel, food, company, or entertainment, they can be deployed for long tenures (Kuhn et al. 2020, 5). As such, Saildrones epitomize elemental prosthesis, indefinitely widening networks on wave materiality into the sea. This may look like a mathematical rendering of oceanic convolutions, but the data is far from comprehensive and actionable. The Saildrone is elementally embedded. Granted, it is designed not to break or corrode when faced with the sea's forces, but it necessarily flows with sea and wind currents, and is propelled by gusts and powered by solar photons. Despite the enhanced intimacy that drones provide, blue governmentality remains at a distance.

Oceanic abstraction has yet to occur, and if oceanographers had budgetary freedom, it would not. They prefer the research vessel over the drone. "The ship surveys [remain] the gold standard," NOAA oceanographer Dr. Calvin Mordy told me (Mordy, personal communication with author, January 10, 2020). "We really see the ship time as critical, but how do we augment in this ever-changing funding environment, right?" asks Chris Meinig, NOAA director of engineering, alluding to why Saildrones are but one tool in the quiver (Pacific Marine Environmental Laboratory 2018). Other NOAA oceanographers add that Saildrones alone "are not a sufficient replacement for trawl sampling" (De Robertis et al. 2019, 10). "Drones are not the answer," Dr. Mordy bluntly told me. "They augment what we need" (Mordy personal communication 2020). Saildrones compose data that is distinct from that collected from a ship because they stretch senses and are conditioned by the elements differently.

After NOAA's 2017 seal mission, the Saildrones sailed north to investigate the edge of Arctic Sea ice—where it is, how cold it is, and how it absorbs carbon dioxide and atmospheric acid. This oceanic ice edge is "a zone of transition, between opposing mediums, a change array for flow structure in the atmosphere and the ocean, a moving habitat, a site for the propagation of biological processes, and a jaggedly uneven line of retreat as larger order physical oceanographic and atmospheric forcings shift and alter its presence in this rapidly changing region" (Steinberg, Kristoffersen, and Shake 2020, 99). During the Saildrone deployment, 2017–19, the Bering Sea warmed precipitously. Ice cover disappeared across the region during winter. This and other factors will likely permanently transform local hunting, industrial fishing, and shipping (Huntington et al. 2020, 342). Cold-water fish like walleye pollock and Pacific cod (*Gadus macrocephalus*) have moved north; Pacific salmon (*Oncorhynchus*), preferring warm waters, migrated south, and NOAA marine biologists are aware of this detail.

The Bering Sea is an ice-mediated system. "As we move forward in the next 20–30 years, we expect there to be less and less ice. So, what does that mean in terms of not just the ecosystem, but how do we change our approach to monitoring that system? What kind of tools do we need to bring in to do that?" (Mordy personal communication 2020). Kuhn told me, "Last year was what we predicted in our models for 50 years from now. And now we see it. So now I don't know what's going to happen" (Kuhn personal communication 2020). The seal scientists understand the recession of Arctic ice but are unwilling, at this critical juncture, to speculate on its effects on seal predation in scientific publications. Without the freedom of the activist science of Ocean Alliance in chapter 2 (as ineffective as it was) nor the adroit activism in chapters 3 and 4, this science is incremental and conservative.

After recapturing the seals and recovering the Crittercams, Kuhn (2017) reflects while reviewing the video back in Seattle, Washington, at NOAA headquarters: "We get to enjoy some time each day swimming under the surface of the Bering Sea checking out the world from a fur seal's point of view!" (figure 5.5). The seal-produced videos display how the seals dive deeper for older, larger pollock; how they catch larger fish near the bottom, smaller fish in large schools, and fish hiding in jellyfish tentacles; and how the seals cohabitate with brittle stars, diving seabirds, and other seals. From the videos, the scientists can identify the pursued fish and estimate the size, age, and thus the nutrient richness of the prey. Linking the video information to the Saildrone's echo-sound data, scientists examine how feeding changes relative to fish availability. Not the nature realism of the activists in the previous two chapters, logical positivism dominates this interior milieu. An energy estimate is made as the scientists correlate the effort expended in tiring pursuit of prey and the size and caloric content of that prey (Kuhn et al. 2020). Due to the intimacy, the seals are conceptualized as energy stores.

Once the seal and Saildrone data has been analyzed, NOAA scientists publish papers that have little chance to improve seal flourishing. They do not intend to do so. This is, after all, "basic science," as they regularly reminded me (Mordy personal communication 2020). The studies published from the research analyze the feasibility of using the Saildrone to follow foraging seals (Kuhn et al. 2020), conclude that the pollock distribution influences seal diving behavior (Mordy et al. 2017), and present the strengths (endurance) and limitations (species identification) of the Saildrone (De Robertis et al. 2019). Reports are submitted for the fisheries management plans that omit the more pressing evidence of the challenges that the seals face in a warming sea with relocating prey.

NOAA has a mandate to abide by US federal environmental regulations and to ensure that the fishing that they endorse does not violate endangered species laws. I do not doubt their compliance. Seal existence is a concern for NOAA, but not a priority. Considering their economic value, it is no surprise that the fishing industries are privileged in allocating the marine resource of pollock. So, while NOAA's annual pollock stock assessment mentions the seals' fluctuating populations and their dependency on pollock, seals are not discussed in NOAA's ecosystem status report, which provides a summary of the allocation for fisheries

(NOAA 2019; Ianelli, Barbeaux, and McKelvey 2020). These summative and largely quantitative reports are governed by the expectations of scientific objectivity, brevity, and other literary conventions of the managerial genre. And yet, possibilities abound for leveraging the insights that relative technicity affords to rebalance reciprocity and summon the existential politics of scientific narrativity. The awesome immediacy of the Blue Anthropocene calls for new conservation technologies as well as representational strategies.

STORYING

The human brain is wired for stories. Senses of hearing, sight, and body sensitivity—and their accompanying prosthetic technologies—gather information that is sculpted into textual or audiovisual stories by orators, producers, and the mind of the audience participant. Using videos, neuroscientists discovered that narratives affect our attitudes by causing the release of oxytocin, a hormone that acts like a neurotransmitter and produces feelings of caring, bonding, attachment, and empathy (Zak 2015). These biochemical shifts make possible the forging of new associations that solidify or challenge attitudes, shifting how we act in the world. Within this framework, stories are appreciated (or not) by a reader or viewer's sense making (Ricoeur 1976). Listeners, readers, characters, and narratives are neither stable nor universal, nor are they chaotic and atomic. Making and hearing stories are interpretive acts. Stories mutate, are remixed, travel unintended networks, and can become something other than intended (van der Leeuw 2020). Indeterminate in outcome, the trajectory of stories through brains, bodies, and representational technologies may lead, however circuitously, toward care.

Stories can be powerful engines of change that reflect and propel historical acts. The underwater storytelling of Jacques Cousteau of France, Valerie and Roy Taylor in Australia, Stan Waterman in the United States, Hans Hass in Germany, and David Attenborough in the United Kingdom offers viewers emotive tales of biodiversity, evolution, and struggle. Cousteau's conservation began with an elemental curiosity. When asked why he was obsessed with the ocean, he said that, as a boy, he "was fascinated by the fluid element capable of supporting very large ships. I could not understand very well how it did it." "But you could have learned all of that by visiting museums and walking about on the surface. What did you get by diving?" the interviewer asked, and heard the retort, "If you read a book about lovemaking it is not the same" (Garbus 2021). The films and television programs produced by these marine storytellers had an immeasurable impact on ocean conservation.

Absent Earth-conscious billionaires buying forests or patrolling marine parks with private navies, conservation in a democracy requires public support. This means translating science and scientific fieldwork into evocative media. Internet video, feature-length films, streaming television, and social media offer platforms for conservation narratives. The typical idea is that stories offered by these media can generate emotional and rational interest in the moral and ecological importance of conservation. But many environmental stories fall on deaf ears. After decades of politically engaged documentaries and few successes, media studies scholars are right to question the efficacy of advocacy cinema (Gaines 1999). The few that are successful are rarely scientific. Conversely, scientists rarely succeed in popularizing their findings. In the pursuit of an equitable approach to narrativity, I offer a conceptualization of storying that might bring together the methods of science and animal worlds.

Animal ethnographers Thom van Dooren and Deborah Bird Rose name science-inspired narrating "storying"—an "ethical work" that requires "responsible thinking" in narrative construction (Van Dooren 2020, 1–2; Van Dooren and Rose 2016, 104). Storying is a multispecies collaboration: "Where one person's or species' knowledge stops, someone else's knowledge picks up the story" (Rose 2013, 104). Environmental stories may ground values, impact decisions, inform actions, and encourage or delay conservation policies. Regardless of their ultimate impact on conservation, storying is a cosmopolitical way for scientists to work well with wildlife. "New animisms and new intimacies," writes science and technology scholar Kath Weston, are generated through animal sensing systems that "thread their way through these alternate stories" (2017, 4). Storying summons the scientist not to erase the artifacts of intimacy, but to fold them into a more complete depiction of the relationships behind the data (Latimer and Miele 2013, 13). Stories dignify multispecies intimacies. For the seals this might mean describing the ordeal they underwent in donating their blood and predation maps. It could incorporate the abstract nonhuman sounds and sights of the seal's point of view, hearing, and embodiment. A forceful argument for the guardianship of sea ice and pollock may be woven into this seal story. The call is to make the entanglements explicit. Such seal stories would be reciprocities and collaborations.

Stories are reciprocities that scientists might engage in as trade for the data they procure from animals. Bringing together the relational thinking of both new materialism and Indigenous scholarship, education theorists Jerry Lee Rosiek, Jimmy Snyder, and Scott L. Pratt offer "an ethics of reciprocity with non-human agents" (2020, 340). This would imply that, when seeking insight into some part of the world, an inquirer would not be entitled to that knowledge. Instead, they would need to consider what they are giving back to the agents coconstituted in the inquiry and the broader network of relations in which the encounter is nested. What is reciprocally given might include service to purposes other than one's own or symbolic gestures that acknowledge interdependence (Rosiek et al. 2020, 340).

Stories are not innocent. In the past, scientific narratives legitimized colonialism, racism, sexism, and anthropocentrism. But today, the global hegemony, driven by carbon capitalism, is unthreatened but nevertheless under siege from science, which demostrates that carbon-based fuels contribute to acidifying and warming the ocean. Climate science offers a "new narrative [that] does not serve the powerful but exposes their absolute failure," writes political theorist Clive Hamilton (2017, 79). Climate science proves that fossil fuel industries cause Earth-warming emissions. In this manner, like Indigenous story work, scientific story work may become a methodology of social and scholarly critique initiated by empirical observations and enacted through narrative performance (Archibald, Lee-Morgan, and De Santolo 2019). Moving toward blue governmentality, we need more-than-humans to make sense-able decisions. As anthropologist Cymene Howe writes, "Sensing, feeling, and knowing the massive yet enigmatic processes known as 'climate change' means interpreting the world through human capacities but not necessarily through humans alone" (2019a, 9). As a new methodology that produces robust data streams, scientific drone work in collaboration with instrumented animals could provoke oceanographers to craft new adversarial narratives of struggle and survival. Whether we listen and respond to these stories is itself another story.

Storying is one narrative technique that has the potential to inform political decision-making. This is not to express a naive understanding of how science informs stories and impacts politics. There is no direct and "deterministic relationship between the production and use of knowledge" (Turnhout et al. 2013, 155). It is a fallacious faith in modernity that assumes that effective science communication to an open-minded and politically active public is all that is required for the protection of biodiversity or reduction of carbon emissions. It does not work that way. As studies of the political action that follows the dire findings of climate science have shown, action is not always the outcome of better data (Hulme 2009).

Storying is a different methodology. By drawing from but not being subservient to the quantification provided by drones and other instruments and methodologies, storying is a narrative method that incorporates the manifold technical, elemental, human, and other animal influences into fuller, thicker depictions of science, its consequences, and its political potentials. Effective action to protect biodiversity "does not necessarily require precise information" (Turnhout et al. 2013, 155). It is not an approach reductive to quantity and commodity, but a mode of temporal, structural, and character elaboration. Storying incorporates the qualitative and quantitative into a depiction of the "messy realities of and dynamic interactions between knowledge production and decision making processes" (Turnhout et al. 2013, 155). It is a technique of technicity, elementality, and animality for narrativizing multiple scientific and qualitative understandings of biodiversity and its existentiality.

Instrumented with sensors, monitored by drones, these seals biolog their lives. But what are they saying, and are we listening? Science scholars Bruno Latour and Peter Weibel conceive of ecologies as democracies and speech as the exchange between species. They recognize "the immense complexity involved for any entity—human or nonhuman—to have a voice, to take a stand, to be counted, to be represented, to be connected to others" (Latour and Weibel 2005, 458–59; Latour 2004). Despite the seal's unintentional participation, humans are not partners in a communicative give-and-take. Ecocritic Serpil Oppermann writes about the importance of multispecies stories in the Blue Anthropocene:

> The storied sea today is a hybridizing mix of the Anthropocene dilemmas within which marine creatures play out entwined ecological crises and material intimacies. And, whether they live in the pelagic or benthic zone . . . they want their voices heard, their stories recognized and their attempts to stay alive understood. . . . But that does not mean they are mute; in fact, they tirelessly project a storied existence conveyed in signs, colors, sounds, signaling,

and codes we may or we may not yet fully understand. Language . . . extends through the marine habitats as an expressive aquatic power when communicative/expressive acts among sea creatures become audible. (2019, 453)

Oppermann's aquatic narratology is not anthropocentric. For her conceptualization, language and signs are not only products of humanity (Kohn 2013). In her thinking, symbols permeate the sea and are expressed by all species in their animistic nomenclature. But in their storying, the seals are not voluntary informants. Nor are they mere passive data-gathering "platforms for oceanographic sampling," as some scientists term these reluctant collaborators (Fedak 2004). Nor are they agents on par with the scientists. The relationship is asymmetrical. The seals give, the humans take. This asymmetry is more problematic than the painful drawing of blood. It is intergenerational trauma for the future.

It could be different. Aided by the drone, sensors, and a sympathetic scribe, the seals could participate in their own salvation. A deeper dive into one of the instruments attached to the seals, the Crittercam, provides an opportunity for a more nuanced understanding of the possibilities of the storied sea, one in which the seal can shift from being *notworked* (that is, intentionally not linked) to networked. It shows what is necessary in affirmative blue governmentalities: better technologies and better storytelling, but also more intimate interactions between drones, scientists, and animals.

For media activists, mobile video cameras provide citizen journalists the means to express themselves, document suppression and resistance, and hold the powerful accountable. Similarly, it is claimed that Crittercams "give the animals the tools to show us directly what's important to them in their life histories, to take the guesswork out of it" (Marshall 2011). Crittercammed seals are producers of a kind of user-generated content (UGC) it seems, once so celebrated in the Web 2.0 era (2005–10) as a form of viewer empowerment (Fish and Srinivasan 2012). Seal-generated content crosses the species barrier from seals to humans via the proliferation of networks. In this techno-ideal, media democratization goes underwater and for multiple species. This ideology of empowerment through participation in media assemblages, be they video, satellite, cable, the internet, drones, or Crittercams, is a central argument of a genre of technoliberalism, which is defined by an idealistic belief that networked technologies provide long-term social benefits—in this case, species conservation, interspecies empathy, eco-entertainment, and scientific data—while also being economically self-sustaining (Fish 2017). Crittercams empower seals while producing profitable entertainment and scientific data, so the discourse goes. The seeming contradiction between "do-good" social liberalism and "make-money" economic liberalism is ameliorated under the header of technoliberalism. With sensing technologies, seal bodies can be both better seen and better capitalized by technologies.

After viewing National Geographic television programming constructed from Crittercam footage collected on the backs of green turtles (*Chelonia mydas*), emperor penguins (*Aptenodytes forsteri*), and humpback whales (*Megaptera novaeangliae*), and hearing promises of human self-transcendence and animal self-representation, Haraway felt as if she were "back in some version of consciousness-raising groups and film projects from the 1970s women's liberation movement" (2008, 252). A return to the ego-dissolving dreams of that period is appropriate. The precedents for the internet, citizen journalism, UGC, and computer-aided augmentation also emerged from West Coast American countercultures in this era (Turner 2006). Many of those projects in consciousness-raising, guerilla television, and activist empowerment moved from political to economic objectives. With the universality of the network you could do (socially) good and (economically) well, without contradiction. Thus, the idea that Crittercams empower seals is technoliberalism that has gone transspecies and para-elemental.

Instrumented seals and Saildrones do enable strange data intimacies, but this is short of transubstantiation and seal empowerment, despite claims to the contrary. "Working with autonomous technologies is allowing us to collect more data than we ever have before. We're swimming in it!" Kuhn (2017) writes (figure 5.5). After a dive on the research submarine *Alvin*, the vessel operators similarly claimed, "we are merging with our data" (Helmreich 2007, 630). Here the "oceanspace" functions as a "virtual reality through which the appropriately cyborg subject might swim" (Helmreich 2007, 630). This self-transcendence and interspecies communication is, of course, illusory. "Oceanographers do not just merge with their data," Helmreich admits. "Submarines do not just dive in unstructured space. And anthropologists do not just soak up culture" (2007, 631). Drones mediate data intimacy, but individuals remain individuals. Oceans and cultures do not become oceancultures but ocean/cultures—parallel but not hybridized worlds.

FIGURE 5.5. Actual "swimming with the data." Video view from diving and hunting instrumented seal. Jellyfish and pollock in the background. Source: NOAA Fisheries.

Drones have not replaced the work of creeping up on and instrumenting seals. Scientists still capture seals, take fluid samples, and separate mothers from pups. Nor have drones replaced the research vessel. Data intimacies are made possible by sensor systems such as *Alvin*, Saildrones, and Crittercams. Ice, ocean, wind, and sun—and the contingencies of more-than-human intimacies—continue to make this research full of the friction it was once thought would disappear with ocean drones (Lehman 2017). The vials filled with seal blood are only the most obvious medium. Oceans and drones mediate this data intimacy. This ordeal exposes how, given that human overfishing and climate-induced ice melting likely harm seals, we can think about possible responses. While Haraway (2012) advocates for "response-ability," the ability to respond to the responsibilities of intimacy, Howe (2019a, 3) calls for "sense-ability," the sensibility to reciprocate that is solicited from the human sensor. Scientists' sense-ability is needed to defend flourishing in the Blue Anthropocene. The alternative is worse.

EXTINCTION MEDIA

The absence of sense-able stories is silence. This instrumented quietude is extinction media, the notworking of networked communication. This theoretical construction draws from critical internet studies that explore how power circulates—or does not—in networks. Similarly, internet scholars Geert Lovink and Joanna Richardson's theory of "sovereign media" is uniquely equipped to theorize an animal's position within networks of science and technology with few nodes. "Sovereign media" comes from a curiosity about the existence of orphaned, neglected, and unappreciated websites. They exist but "demand no attention" (Lovink and Richardson 2001, 1). Insofar as communication is collegial and monologues are not communication, sovereign media are "media without message" (Lovink and Richardson 2001, 2). They are ends in themselves—full stop. Considering the rich yet unappreciated data that streams from instrumented animals and networked nature, sovereign media is an apt concept.

Sovereign media typify "notworking," a play on "networking" without effect (Lovink 2005). Notworking invites new perspectives on animal labor in the surveilled sea. Like a workers' strike, refusal can be a political act. But these animals are working against the drag of attached cameras, their bodies bled, and their pups chased. The data they trigger circulates in small, fractured loops of scientific instruments, computer networks, and publications. They help scientific careers and complete mandated reports. This seal data, however, is not connected to political action and storying, and it thus fails to inspire regulations designed to encourage prospering. And thus, "from an activist point of view," not connecting is "completely useless," Lovink confessed (Boler 2008, 128). This politics of absence is extinction media, a sounding without a ping. By hailing and being tracked by the Saildrone, the seals create signals that govern the movement of robot-like drones. But while the instruments on the seals and the Saildrone communicate with each other, the feedback loop potentially affecting their livelihood does not integrate the seals in its circuit. Despite the sensors, cameras, and drones monitoring the animals, theirs is not a cybernetic system of self-regulation, maintenance, and sustainability. For the circuit to be complete and the notwork to become a network, conservation plans and the politics of preservation would need to become dynamic actants. Without this, this Saildrone-seal research remains "basic research."

The topical corrections—the buried blogs that the oceanographers and engineers write and that NOAA and Saildrone Inc. dispense, for instance—are insufficient. The underappreciated scientific papers and absent recommendations to protect the species are not reciprocities that give voice, nor are they dignified reciprocal exchanges. This is not "tactical media"—artful engagements within networked public spheres (Srinivasan and Fish 2011). This is extinction media wherein sound and light waves coming from the seals do not reach their destination.

Seals generate content; they involuntarily assist in the ocean "becoming computational" (Gabrys 2016, 95). They are chased and netted. The seals undergo disruptive capture and blood extraction. They are tagged with an apparatus for surveillance. For this donation, what is given to the seals? Where is the love? Anthropologist Eben Kirksey inspires such a query by defining "love" between "ensembles of selves [as] . . . associations of entangled agents involved in relations of reciprocity and accountability" (2019, 207). Despite the intimate and ephemeral relationship, there is little love between seals and humans. Reciprocal listening and responding are absent. Through Crittercams, the seals are seen but not heard. The absence of effects in this circulation of affect amounts to a politics of negation. At best this is muffled voicefulness, a repression told through murmurations. In the seal's sovereign media is the spoken but unheard utterances of extinction media.

Critical theorist Jodi Dean expands on the notworking of extinction media or circulation without dialogue. Notworking nature, "far from enhancing democratic governance or resistance, results in precisely the opposite" (Dean 2005, 53). Postpolitics is an ideology that enacts global consensus and compromise through technologies. Dean disagrees with postpolitics because in it "real antagonism or dissent is foreclosed" (2005, 56). In this manner, the seal project is postpolitical in its silence; it is circulation without political potency. This "basic research" is scholarship without a practical output. Its findings do not inform policy. NOAA scientist Mordy told me, "Our work is not totally [about] providing metrics for management. It's much broader than that" (Mordy personal communication 2020). It is not enough to speak, or even to be heard by a select few. Democratic political change occurs through agonism, deliberation, contestation, reform, and action. The circulation of "content" on networks is meaningless without noise, friction, and argumentation within the circuit. This starts with body-extending technologies transgressing elements and proceeds through honoring cross-species intimacies with stories. Short of this is extinction media, the notworking of sovereign media at the existential edge of multispecies survival.

Saildrone data has been analyzed, but its findings have not been integrated into a convincing story capable of influencing fisheries management. Subsistence hunting and survival amid a climate crisis—these are obvious messages. NOAA scientists see the signs of these messages but are reluctant to include these interpretations of the seal signals in their peer-reviewed publications, which have the potential to impact US federal fishing policy. They fail to compose a compelling narrative about seal lifeways. Instead, seal Crittercam videos are uploaded to the federal government's underappreciated website and accompany presentations to small audiences at academic conferences. In this manner, the seal's message of crisis morphs into content with a thin circulation. Drones generate evocative content. As elemental technologies they can afford attunements with species. They are intimate technologies that bring humans and nonhumans together. But they have yet to become existential technologies—instruments for the preservation of life.

If they are to be heard, the seals must speak in the language of the human scientists. As literary scholar Gayatri Spivak (1988) expressed, marginalized communities must pronounce in the terms and on the platforms governed by dominant elites. And yet the scientists must be cautious to not become what Haraway (1997) describes as scientific ventriloquists, speakers for instead of enablers of speech. This will require hearing before uttering. Aided by human and technological proxies, animals might become actors in their own histories. In this spirit of active listening and reciprocity, seals, sensors, cameras, and pollock could combine to narrate a unique story. Storying can begin with scientific writing where positivistic publications end, potentially influencing publics and policy. A story the scientists might tell is about precarious survival in a changing sea. Bering Sea ice is migrating northward—and with it the seals' food. The pollock are overfished by humans. A population of pollock in the Bering Sea has already been decimated in what one historian describes as "the most spectacular fishery collapse in North American history" (Bailey 2011). Despite the warming sea and receding ice, which threaten the pollock and with them the seals, the pollock catch quotas set by the US government remain high, making billions of dollars for fishing industries (Bailey et al. 2000). Regulatory intervention by NOAA to limit this catch for the seals has yet to occur, negating the existential reciprocity that data intimacy demands.

Along with relative technicity and elementality, storying is but one factor that contributes to the efficacy of blue governmentality. As studies have shown, there are limits to the empathy that compelling environmental narratives engender (van der Leeuw 2020). Many are already deluged by professional and user-generated content and probably do not have time for additional animal streaming—a quarter of American children already spend eight hours on screens a day (Hartshorne et al. 2021). But unlike relative technicity, which is driven by evolution, and elementality, with its unpredictable physics, storying is an activity that humans have agency over. As all that screen time suggests, we are biochemically hardwired to be captivated by stories, and story makers traditionally and currently hold positions of influence in human life. Like technology, storying is something we do well. This chapter suggests how novel sensing technologies combined with an ethics of reciprocities might push scientists to do it better.

TELLING SEAL STORIES

My argument is not that NOAA is violating regulation nor that basic research is unethical. Rather, my point is more philosophical and speculative of an affirmative future of multispecies reciprocities. Drones afford scientists a proximity that generates interspecies intimacies and in turn requires a give and take. Although it is obvious that drones provide higher resolution and granularity of biological data for scientists, drone oceanographers could better reciprocate to ensure marine species' survival. I am not concerned with reforming ethical guidelines for the treatment of marine mammals to stipulate how not to harm animals when pursuing, tagging, restraining, and drugging them (Society for Marine Mammalogy n.d.). What I am proposing is a commitment to flourishing that becomes pervasive within and alongside scientific publications. Within the bounds of veracity and its strict textual limitations, this existential imperative could be smuggled into scientific reports through narrative conventions. Storying could improve the public understanding of scientific insights and invite better care for marine species. One could do worse than look to the wealth of compelling writing that draws from scientific studies, is conscious of the constitutive role played by technology and the elements, and is empathic toward marine species and their survival as models (e.g., Giggs 2020; Schweitzer 2014).

The intimacies produced by Crittercam are magnified by the other sensing technologies that constellate around the seal. The Saildrone allows closer, quieter proximity than the research vessel; the dive trackers provide precise longitudinal and volumetric data; seal blood draws require capture and struggle between hands and flippers. Technological sophistication, scientific ambition, federal funds—and empathy and care—are needed to hear and visualize the seal and pollock's undersea stories. The threats to their livelihood become legible in videos of seals chasing fish, are triangulated by drone-made sonar depictions of pollock schools, and are made human sensible by oceanographers who might inform American fisheries policy. In these posthuman collaborations, the seals are agents despite their involuntary participation and the existential irrelevance of their input to fisheries policy.

In the future, Saildrones and their kin will become even more ubiquitous and responsive. Instead of the GPS location data bouncing off an Argos buoy and a satellite, the seal's location could go directly to the Saildrone, decreasing the time delay in the Saildrone's response to the seal's movements. Ubiquitous trackers on other species or monitoring other habitats could inform the Saildrone's awareness of the seal's location (Kuhn et al. 2020, 6). These future innovations may require more sensors, more Saildrones, and a more wired sea. For now, the seal diving record, sea video, Saildrone fish records, and pup health will make known not only how important pollock are for seals, but also how important technologies are to our understanding of wildlife. These seals are not yet entrapped by conservation management; they are still semi-autonomous from the controls that these sensors might actuate. This is beneficial for their relative autonomy, but it threatens their long-term survival, which may require policies informed by sensors for continued thriving.

Seal data emerges from a transduction, a rendering of sound and raw data into malleable waveforms and words. Once rendered, such texts can become stories of survival that call attention to struggle and make possible the care and control that resist extinction. Without these stories, the seals' voices are unheard. Their flourishing is unnarrated. Without these science-inspired stories influencing publics and informing management, seal vivacity remains precarious. This is vitality without the multispecies justice they deserve for the data the seals provided. So, while sea-surface drones track the seals' movements, these conservation technologies do not enact a version of blue governmentality—a control of elementally situated seal biopower. Instead, we witness a sophisticated, multimodal scientific methodology without existential exchange.

I offer storying as a practice that builds persuasive narratives to honor our multispecies interdependencies. To expand this flourishing, insights from drone oceanography point to better leveraging the present revolution in robotics—exemplified by the drone—to address this moment of oceanic demise. For oceanographers, ocean scholars, and ocean-dependent beings, responding affirmatively to the challenges of the present will require more than drones, however. Technology, storying, public support, and legislation are a suite of forces within blue governmentality. Moving with technologies in acts of storying elevates the struggles of marine life, valuing the complex web of humans and nonhumans in the Blue Anthropocene.

6

CRASHING
FALLING DRONES
AND ABANDONED
TERN COLONIES

CORAL CRASHING

In 2018, I traveled to Bunaken Marine National Park in Northern Sulawesi, Indonesia, in the heart of the Coral Triangle—the global center of marine biodiversity, which includes the Philippines, most of Indonesia, and Papua New Guinea—to see how drones might be useful in identifying coral bleaching (figure 6.1). Over several days we collected over five hundred high-definition still images of the reef and stitched these shots into a single, massive, highly accurate map of bleached coral. In response to a warming ocean, coral expels its nutrient-providing algae, leaving white skeletons in turquoise waters. The dying coral is heartbreaking but easy to identify from the air.

Near the end of the project, the drone crashed.

It was unspectacular. After avoiding the spears thrown by young octopus hunters, the little DJI Mavic Air met an unceremonious end, falling into the ocean from a rickety pier. Attempts were made to rinse the salt from its components in the least brackish water we could find. But the salt

FIGURE 6.1. Coral reef at Bunaken Marine National Park, North Sulawesi, Indonesia. Photo by author.

calcified in the motherboard, frying it. The drone manufacturer shipped us another drone. Insurance paid, as if the first drone was merely disposable. And I guess it was, because it was neither recyclable nor reusable. The engine of this dissipative economy is planned obsolescence. Our drones' copper, silver, gold, palladium, lithium, and indium were slated for a shallow landfill on this island.

The high-resolution images we built corroborate evidence of coral bleaching from satellite imagery and other data signals, and together these studies form a longitudinal depiction of coral bleaching around Bunaken Island (Ampou et al. 2018). Our experiment in drone mapping contributes a historic slice of the reef with a resolution better than a satellite. Our orthomosaic map adds to a study of the impact of climate change and ocean acidification on coral reefs. In previous studies, drones and underwater cameras created "sliced ecosystems," stratigraphic depictions that illustrate how excessive tourism, overfishing, and other activities harm this fragile area (Rekittke and Ninsalam 2016). On Bunaken Island ecology overlaps with archaeology to uncover the elemental layers of intra-action between people, technologies, and ecologies. The drone is an organ at the edge: the end of the drone supply line, the boundary of digital connectivity at the shoreline, the border separating the earth and sea, and the space that separates functioning drones and faltering ecosystems.

As elevated networked sensors, drones work the atmospheric edge of the global information infrastructure, expanding the internet into horizontal and vertical space. This relative technicity aligns with the thinking of media scholar Fredrich Kittler (2009) and his materialistic ontology of technical media. According to Kittler, philosophy does not merely comment on phenomena; its insights are themselves determined by the means of philosophical production—alphabetical systems, writing utensils, and the furniture of printing presses, for example. In a Kittlerian framework, what we think of as a drone is structured by what the drone allows us to do, see, and collect from the elements. Drones are aeromobile, optical, and networked sensor systems that are manufactured to eclipse distance, manifest commands, receive data, and store information. Saildrones are designed to go where we cannot: they can travel conveniently into the atmosphere or, as we saw in the previous chapter, across the surface of the sea. The drone's ontology is formed through mutually reinforcing atmospheric constraints, ecological possibilities, multispecies interactions, and human-generated ambitions. Understanding the potentials of computing technologies requires focusing on the capacities of transmission to eclipse speed and data storage to resist the decay of entropy.

Kittler connects technical media's properties and potentials to practices such as writing, coding, and recording, but also to destructive efforts such as erasure, editing, and recoding. He sees the physical similarities between using creative technologies of media production (shooting a picture) and deadly weapons of war (shooting a gun). Kittler's techno-ontological emphasis on transmission, physical media, and distance aligns with the mechanical operations of drones and invites us to meditate on edges, limits, boundaries, and breaks. My drone's technicity on that fateful afternoon in Indonesia, compromised as it was after falling into the ocean, shows the limits of my ability to document coral bleaching. This drone's technicity is found not only in its capacity to fly and see but is also evident in its demise.

The atmosphere and the liquid ocean are volumes with material consistencies that lift drones, giving them the resistance they need to propel across space. These elements also disrupt drone lift and movement, challenging its trajectory with gusts, swell, salt, sand, ice, and storms. One unfortunate outcome of this elementality is a crashed drone. Hull damaged, propellers cracked, connection lost, circuit boards corroded, the drone sunk below the surface or evaporated into the horizon—the crash has many paths but only one shared fate. Difficult to recycle and unrepairable by design, its rare earth metals in a landfill, the drone has

FIGURE 6.2. Drone view of erupting Agung volcano, Bali, Indonesia, shortly before crash. Source: Video still from documentary *Crash Theory* (Fish 2019a).

an end that is a fitting display of modernity's drive for fast movement and rapid consumption.

Over the course of the fieldwork for this book, I participated in many drone crashes. In Indonesia, we crashed as we surveyed the erupting Mount Agung volcano in Bali (figure 6.2) and while exploring the undersea fiber-optical system into and out of Iceland (figure 6.3). Because of their regular occurrence, the work taken to avoid them, and the morbid fascination they hold, crashes are not the meaningless noise of drone culture, but rather a signal that perseveres through destruction.

As I discovered in my discussions with drone pilots around the world, crashes are common. The hardware and software are not perfect. Flying is new for many people. The exhilaration of flight is intoxicating and, therefore, distracting. A cottage industry of drone crash videos on YouTube has blossomed along with the proliferation of comments of cringe, schadenfreude, and shame. Given the increase in drone use, it is likely that crashing drones will continue to threaten people, animals, and property. Safety will improve, but today the crash is part of drone culture. Technological innovation, experimental practice, and accidents are intertwined. As cultural theorist Paul Virilio (1999, 89) famously quipped, "when you invent the plane you also invent the plane crash." Not only is the drone crash invented with the drone's invention, but the drone and its crash have become connected to more-than-human processes.

FIGURE 6.3. Drone crashing in Iceland while mapping undersea internet cables. Source: Video still from documentary *Crash Theory* (Fish 2019a).

Crashes reveal alien agencies. Consider another drone crash I had the misfortune to contribute to. It was 2015 and I was in Landeyjarsandur, southern Iceland, mapping the undersea internet cables into and out of the island from Europe, with cultural geographer Bradley L. Garrett. On one flight, our Inspire DJI drone sped away from the cable station at an alarming speed. It ignored the five-hundred-meter limitation that can be programmed into it. Instead of warning us that it had reached its spatial limit and would soon be returning to its point of departure, the drone continued to trace the undersea internet cable beyond the expanse of black sand and out into the North Atlantic Ocean. This was a liberating experience, but also terrifying. The euphoria quickly ended when we realized we were about to lose the drone in the sea. For whatever reason, we were able to regain control over the drone. Turning it around, we brought it back to shore. But the drone's experiment in temporary sovereignty had drained its battery, and it started a descent—not a fall but a smooth plunge as the four propellers spent their last spins to drop the drone in one of the grass patches growing out of the stacks of hay that protected the internet cable from the shifting sea dunes (figure 6.3) (Fish, Garrett, and Case 2017a, 2017b).

Coming down from our panicked and dazed height, we asked why the drone did what it did. Was this augmented flight at Landeyjarsandur a failure or an opening? We were reminded of urbanist Stephen Graham, who wrote, "Moments of stasis and disrupted flow [can be] a powerful means of revealing the politics of the normal circulations of globalizing urban life" (2009, 3). Perhaps it was the conductivity of the basalt in the black sand. Maybe it had something to do with the way the drone and the electromagnetic field of the cable landing site interacted. Who knows? But this drone getting away from us was one of our first experiences of exosomatic extensionality. Losing control was rousing and terrifying.

When meeting each other, seasoned pilots talk about crashes—how often, how bad, who was to blame (the operator, manufacturer, or environment). With a mix of danger, excitement, and machismo—most pilots I have encountered are men—this talk of crashes resembles the stories war correspondents tell each other as a way of creating an informal social ranking (Pedelty 1995). I too participated in this game of bravado, attempting to leverage my mistakes to gain credibility with project participants. In this manner, crashes became technological sacrifices in initiation rituals in fieldwork (Geertz 1973). Despite substantial efforts to avoid crashes, they still occur for even the most prudent professionals.

Crash avoidance is integral to drone work—training sessions, preflight checklists, the monitoring of the weather and wind, meticulous care for gear, close attention to takeoff and landing spots, the application of automatic flight corrections, compliance with regulations. Like elementality itself, the crash is not a bug but a feature of drone culture. It provides ontological insights into the nature of the drone, its flying, and the objects it explores. Crashes happen and with real consequences for drones, pilots, and the living beings around the crash's epicenter.

Metaphor offers an expedient interpretation of the crash. An average hermeneutics might be that the drone crash symbolizes, for instance, our hopes of technological progress, which are dashed into the corrosive ocean. A persuasive symbol might be that the crashing drone represents ecological demise. This kind of representationalism is anthropocentric, however. It contradicts this book's advocacy for posthuman elevations of the influential agencies (without intentionality) of technologies and elements. Such a metaphorical approach exclusively engages human-based symbols and thus human-focused concerns. The crash is more than metaphorically relevant for understanding the political impacts of recent technologies. To think more clearly about how the crashing drone impacts marine species, we need a concept that physically and existentially connects the two—a theory that examines how seeing technologies and the subjects they explore are interrelated.

The work of feminist quantum physicist Karen Barad (2007) provides a way of understanding the cocreation of optical instruments like drones and that which they explore. Her "agential realism" argues that technological apparatuses like optical scientific instruments and their observers do not interact but rather intra-act, forming a material-discursive phenomenon in which technologies, scientists, and that which they encounter—internet cables or volcanoes, whales or terns, photons or electrons, photons reflecting off whale skin and bird feathers—mutually influence each other.

Mediating the physics of waves such as light and electricity, the drone is a quintessential apparatus, in the Baradian sense. Drones spread from the bodies of scientists and activists, inviting ocean animals into meaningful and, ideally, ethically formed phenomena. Drones traffic in quanta—photons and electrons that travel the electromagnetic spectrum. Drones are one of several relata, or related subjects, in relationship with humans and nonhuman others. Drones become alongside these others. Together, these things-in-process—including drone apparatuses, photon and electron quanta, and the living and nonliving others—form a drone phenomenon. The evidence of the phenomenon is a diffraction, a material or visual artifact of overlapping dissonances. The crash is a poignant diffraction, a reminder of disentanglement and the separation of nature and culture.

Specifically, the crash in the drone phenomenon makes entropy evident. Endangered species, whose populations are precipitously dropping, and the falling drone are both shifting from complexity to a simpler, higher-entropy phase state. The conservation drone, when not crashing, attempts to reverse entropy by repairing populations. Breakage-begetting crashes reinforce this ethics of repair (Jackson 2014). Acts of repair involve technologies in the management or programming of natural worlds (Gabrys 2016). This process can be seen as the "reworlding" of blue governmentality, a form of environmental justice that is centered on multispecies control and care. (Haraway 2016, 40).

This chapter's two case studies explore legal contingency within blue governmentality through the specter of the crash. Blue governmentality, you will recall, is an effort at affirmative control of life limited by the elements, human sociality, animality, and technological fallibility. Taking an agential realistic approach, this chapter examines what two crashes in the United States mean for affected animals, the failing human pilots, and the tenuous efficacy of blue governmentality. These collisions diffract prosecutability, the ability to be apprehended and charged with a crime.

In the Puget Sound of Washington State, the laws designed to guard against crashes of vehicles and orcas are one contingency that is diffracted by the threat of a crashing drone. On the coastal dunes of Southern California, a life-decimating drone crash exposed the tenuous existence of nesting elegant terns (*Thalasseus elegans*) and the capacities of police to enforce their prospering. These case studies exhibit the legal porosity of blue governmentality.

COLONY COLLAPSE

On May 13, 2021, Peter Knapp, wildlife monitor with the California Department of Fish and Wildlife, was out on his daily survey of the elegant tern population in Bolsa Chica Ecological Reserve, a land of open water, coastal dunes, and salt and freshwater marshes in Southern California. Bolsa Chica is one of only three nesting sites in California for the bird (Firozi 2021). Three thousand terns had recently arrived, and Knapp was excited to document how the nests were developing. Gusting winds blew over the birds who squatted on clutches of eggs. But today the birds were gone. Usually, the aggressive birds would attack any visitor as a would-be assailant, but Knapp was free to wander into the rookery without being dive bombed by worried parents. Instead of nesting birds, he found a crashed DJI Mavic 2, its exterior coated in shell-white bird feces. Surrounding the mangle were 1,500 to 2,000 cold and dying eggs (figure 6.4). Scared off by the drone's inadvertent physical mimicry of the tern's primary predator, the peregrine falcon (*Falco peregrinus*), the parent terns had abandoned their eggs. A generation gone in a crash.

The elegant tern is loyal to the flock, not the breeding ground. "If some individuals take off for whatever reason," Kate Goodenough, a seabird ecologist at the University of Oklahoma, said, "the majority of the group also takes off and leaves" (Thompson 2021). Melissa Loebl, an environmental scientist who manages the reserve, said, "That's one of the largest losses we've had" (Firozi 2021).

Drones bother animals, birds, and humans. The sound, sight, and idea of a camera in the sky irritates many living beings. For drone aficionados, an important part of the technology's attraction lies in getting closer to animals and seeing them more vividly than they could with the naked eye

FIGURE 6.4. Thousands of elegant tern eggs abandoned in the Bolsa Chica Ecological Reserve. Source: California Department of Fish and Wildlife (Thompson 2021).

or with binoculars. Wildlife viewing is a popular application of drones. Scholars who quantitatively analyzed a collection of wildlife drone videos concluded that the prevalence of "videos with flashy titles alluding to attacks or collisions with wildlife with drones, particularly birds, suggests that these acts have become naturalized" (Rebolo-Ifrán et al. 2019). The harassment of a mother bear and her cub by a drone as they tried to escape up a frozen embankment, only for the cub to repeatedly slide agonizingly down a steep cliff face, went viral (Goodyear 2018). Most terrestrial animals and birds do not like drones. Green monkeys (*Chlorocebus sabaeus*) of West Africa sound the call they make for predatory eagles when a drone passes (Fox 2019). A wedge-tailed eagle took out a drone flying above a wheat farm in Western Australia (Pendergast 2017). Similar tales are numerous. The antagonism between birds of prey and drones was leveraged by Dutch police, who trained eagles to incapacitate their technological competitors (Ong 2017).

Alongside the countless YouTube and TikTok videos of drones crashing into every conceivable object—windows, power poles, trains, boats, hot air balloons, bridges, prisons, oil refineries, oil pipelines, nuclear power plants, airplanes, helicopters, stadiums, the White House lawn, Seattle's Space Needle, and the Japanese prime minister's residence (Dedrone 2019)—there is a video genre of drones chasing or being chased by animals. Cheetahs, chimpanzees, kangaroos, crocodiles, and geese attack drones. Buffalos and other ungulates run and hide from their menace. Likely because the ocean transduces air-based soundwaves into registers that are difficult for aquatic mammals to hear, they appear to respond less to drones than do flying and land animals (Rebolo-Ifrán, Graña Grilli, and Lambertucci 2019). The elegant tern, more committed to flock solidarity than territoriality and offspring, can be added to the list of organisms that do not live well with drones.

On the Thursday following the collapse of the rookery, while reporters for KABC-TV Channel 7 interviewed Loebl and Nick Molsberry, a warden for the state Department of Fish and Wildlife, about the drone problem, a man drove into the parking lot and started to fly a drone toward a neighboring tern colony. "I actually ended the interview, contacted the individual, identified myself, and issued that person a citation right there on the spot," Molsberry said. Later, television cameras captured another drone flying over the reserve; its operator could not be identified (Wigglesworth 2021). "It's ironic," Molsberry said. "Drone owners are attracted by the nesting colonies of birds, and then their actions destroy it." It is doubly ironic because drones have been heralded by ornithologists for their ability to quickly count and map birds, a safer and less expensive alternative to helicopter surveys. The drone, it seems, is integrated into seabird life and death.

The US Federal Aviation Administration (FAA) issued a statement that it was aware of the incident and investigating it. Molsberry is working with the Orange County district attorney's office to figure out how to get a warrant and retrieve video and the flight history from the drone (Wigglesworth 2021). As of this writing, the perpetrator has yet to be found. Without apprehending them and developing a way to protect tern colonies from drones, their coexistence will remain tenuous. To care for terns requires increased control: employing geofencing software that creates no-fly zones for drones over sensitive areas like Bolsa Chica; deploying rangers with antidrone technologies that jam or hijack signals; increasing punitive threats; or cracking the security of suspicious and abandoned drones. None of these options come without damages to human freedom and privacy. But these are sacrifices worth making for tern flourishing.

As a nonanthropocentric theory of biopolitical control in multispecies and posthuman phenomena, blue governmentality pushes us to make this reparation.

This story of precarious survival at Bolsa Chica brings this book's key ideas together around Barad's concepts. The drone is an instrument of relative technicity with both existential and legal consequences. It influences and is influenced by the nonhuman atmosphere and the elegant tern colony. Alongside an irresponsible pilot and the wind that lifted it (and factored into its crash), it forms a phenomenon. The care of the reserve manifests in its capacity to control (or not) technicity—the rogue enactment of movement, vision, and force from afar. As the investigation into the pilot continues, a tern generation is lost, and the possibilities of a blue governmentality that is life affirming recede as the eggs rot, the flock flies from Bolsa Chica, and the drone operator drives away on a windy California evening.

In this brief case, the misflown drone and a generation of elegant terns are entangled in the fate of entropy. If the drone doesn't crash, the birds survive. Legal governance, police investigations, modifications in regulation—these are all downwind from these two codetermining physical states. In a world that is falling apart, our responsibility is to care and repair, and to reverse trajectories toward high entropy whenever possible. This means using drones for conservation and treating drones with the respect to repair they deserve. It demands thoughtful piloting that cares about the living others with whom we are engaging.

Marine protection laws are an integral relata in this drone phenomenon. They enforce compliance with procedures that enfold oceans in human values. The ocean is incompletely humanized by marine laws, however. There is no unmitigated exercise of power in this fluid world of nonhuman agency (without intentionality) and migrant human agency. While such laws bring human procedures and values to ocean worlds, they also segregate humans and their technologies from marine spaces and species. Blue governmentality contributes to a complex drone phenomenon that is moderated by the limited affordances of technicity, the elements, and prosecutibility.

In this chapter's second case study, I investigate another legal aspect of drone crashing. In the Puget Sound, Washington State, a regulatory dispute over piloting drones over iconic black-and-white orcas diffracts the limits of conservation law and, with it, the ceiling of blue governmentality.

My family was considering selling our old seaside property outside of Seattle, Washington, so I volunteered to survey the land with my drone. When we returned to the United States in 2018 from fieldwork in Indonesia, I became aware of the Washington State proposal to ban flying drones over endangered orcas in the Puget Sound.

That year, the Southern Resident orca pod made international news when one mother, Tahlequah, carried her dead calf for seventeen days and thousands of kilometers following what one observer called a funeral "ritual or ceremony" (Crilly 2018). Marine biologists used drones to track Tahlequah and her pod and measure the width of the orcas to predict whether they were pregnant. This experimental yet promising approach revealed a pregnant mother in 2021 (figure 6.5). Every birth matters: this orca population is dwindling, down to seventy-three individuals in early 2021, and researchers are investigating reasons for their demise (Solly 2018). Causes include increasing oil tanker traffic in the Puget Sound, warming water, and viruses from dogs and humans that cross the species barrier. The most likely explanation is that they are starving because of the absence of Chinook salmon (*Oncorhynchus tshawytscha*)—which have been decimated by overfishing, pollution, and the Snake and Columbia River dams, which prevent the fish from returning inland to spawn (Robbins 2018).

While drones are used to monitor orca pregnancies and to collect data about their health, as explored in chapter 2, they are also deemed a problem for the whales. This brings us to the drone photographer Douglas Shih, a wealthy real estate agent from Mercer Island, Washington. On August 15, 2015, along with fifteen other watercrafts, Shih's fifteen-meter-long yacht was following an orca pod when he decided to use a drone to get a vertical view of the family (Shih 2015). While Shih collected evocative footage, this flight violated Washington State law that requires all operators of "crafts and other objects" to stay two hundred meters away from orcas. Shih's drone was twenty meters above the whales when Fish and Wildlife sergeant Russ Mullins spotted it, and for this violation Shih received a fine of US$1,025. Shih hired lawyers and likely spent more than the amount of the citation to appeal it. Eventually, Shih won his appeal on the grounds that the law did not state that "drones" had to remain two hundred meters away—only "objects" (Banse 2016).

Washington State congresspeople responded to this ambiguity, intro-

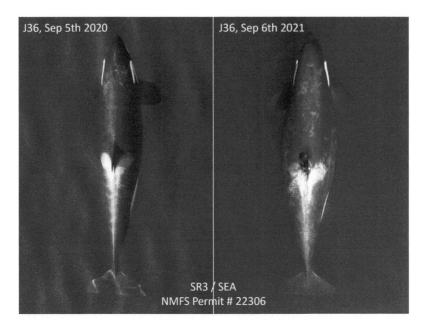

J36, Sep 5th 2020

J36, Sep 6th 2021

SR3 / SEA
NMFS Permit # 22306

FIGURE 6.5. Drone-collected images of an orca's body shape changing during pregnancy. Adult female orca (J36) is pregnant in the image on the right. Images collected by SR3 in September 2020 under NMFS research permit no. 19091 and September 2021 under NMFS permit no. 22306. Source: SR3 2021.

ducing a bill in 2018 that explicitly prohibited "drones" from being closer than two hundred meters from orcas. State senators were confident that it would become law. But Republican state senator Jim Honeyford submitted an amendment that removed all prohibitions against using drones around orcas. In the chamber, Honeyford said, "The orcas are a really big tourism attraction in the Puget Sound. . . . This would allow those tourists who have unmanned aircraft or drones or whatever you want to call them to be able to fly them. They are electric, and they are quiet, and they can take pictures. I believe it would be a great increase in tourism" (Dunagan 2018). Honeyford is opposed to any regulation that might limit the economic potential of the developing drone industry (Honeyford, interview with author, July 15, 2018).

I spoke with US Fish and Wildlife sergeant Russ Mullins, who cited Shih, and we discussed drone threats. Mullins is concerned with crashes. Piloting a yacht is difficult enough without the added challenge of flying an object in the atmosphere. Mullins discussed the monitoring of the orcas. "We'll follow the whales around for ten hours a day." "At what distance?" I asked, surprised. "Well, we usually stand off a little more than the minimum and just observe. But everyone's required to stay two hundred yards away from the whales—that doesn't always happen." "So, ten hours a day for certain periods," I continued. "Well, sometimes, five days a week." "Okay, so we're in a situation where we're almost perpetually monitoring this pod?" I asked, surprised by the commitment to monitoring. "Yes, actually three pods," he admitted (Mullins interview with author, July 1, 2018). Drones are objects that harass orcas and distract humans, causing what Mullins told me is an "arms race" of video technologies competing for the best video of orcas for social media. In the process of filming, orcas, drones, laws, technologies, fetuses, the sea, and the atmosphere are woven into a cycle of life and death. A crash—of drones, boats, or species—was what was to be avoided through the failed 2018 bill. And yet the crash is not a metaphor but rather a materialization of the limits of regulations designed to encourage orca flourishing.

In this orca-drone-law phenomenon, relata include a polluted Puget Sound, emaciated orcas, conservation authorities, and humans in the Washington State legislature debating where drones will be permitted for hobby cinematographers. Crashing drones, crashing boats, and crashing species are knotted together. A drone or seacraft crash diffracts regulation, exhibiting friction between human laws and the survival of sea mammals. This diffraction also reveals humans' dominant position within the conservation drone phenomenon. Human laws enact ethical and partial care in and around orca bodies. In the process, orcas become entrapped, dependent on human laws and surveillance.

As I have demonstrated throughout this book, blue governmentality rarely protects multispecies thriving—at least not yet. Rather, conservation drones crash and their pilots are released without punishment. The drone crash ruptures the promise of blue governmentality. It exposes the ease with which wealthy drone operators avoid prosecution. This is the gap between ocean/cultures: where care meets control, ocean and cultures make contact. In these failures of legal and technical control, they separate. Thus, in their imperfect applications of blue governmentality, the laws against the use of drones above protected marine wildlife, like their crashes, bring together and push apart ocean/cultures. What blue governmentality hopes to achieve is a parallel togetherness of oceans and cultures, a watchful care within ocean/cultures, and a forceful control without.

The drone does not usher in an atmospheric enclosure, a totalizing surveillance of nature, or a militarization of conservation. Drones break, and quite enthusiastically, because of human error, environmental contingencies, and unforeseeable realities. Legal processes around drones are not entirely effective and remain in flux. A theory of the crash can help us understand experimental technologies of control and demonstrate that things are often beyond human control. The absence of control is evident in technological advances of drones that outpace regulations and in the dream that technologies such as drones could safely control ocean/cultures. Science scholar John Law noted that aircraft crashes are messes that "necessarily exceed our capacity to know them" (2004, 6). Perhaps, but the crashes in these two cases diffract prosecutability, an aspect of the uneven application of blue governmentality.

The consequence of human error, complex systems, or faulty organizations (Perrow 1984), accidents are an unavoidable facet of aerospace science (Beck 1992). Sociologist Diane Vaughan's investigation into the crash of NASA's space shuttle *Challenger* in 1986 recognized that despite the known fatal leakages in the rocket's O-rings, the flight proceeded. This is the "normalization of deviance" (Vaughan 1997), an example of the routinization of risk. The *Challenger* explosion showed how aerospace technologies mingle with but also collapse alongside diverse living beings, social systems, and other technologies. Like the *Challenger* disaster, drone crashes diffract the uncertainties of aerospace engineering. Legal consequences follow but are less worrisome concerns in these life-and-death failures.

In aerospace engineering, the constant threat of a crash generates a diffraction that betrays safety and security. According to feminist film theorist Karen Redrobe, the crash in cinema is "a conceptual paradigm of relationality" (2010, 8). It is a diffraction of a phenomenon that includes cars, cameras, producer agencies, spectacles of attractions, and paying viewers' thrills. Collisions "bring difference to the fore within a framework of uncomfortable, sometimes painful, and even fatal, proximity" (Redrobe 2010, 22). The conservation drone crash creates a diffraction that unmasks the fragile, high-stakes relations between technologies of representation, endangered species, and human relata such as policing and legislation. The crash demands a sudden pause and revision. It is an "event" (Badiou 2005), a rupture in the perceived historical trajectory of time for all involved—pilots, drones, and animals. The "crash, as an *event*, cannot be contained, and this is precisely the source of its compelling power" (Crandall 2011, 12). Thus, the crash diffracts evidence of the foibles of field science and legal mechanisms, the inevitable demise of present socio-technical phenomena, and the slippery alignment of oceans and culture.

Two military drone crashes in 2011 illustrate how crashes broaden out and incorporate others from their impact craters. Military and civilian drones are dramatically different technologies in size, capacity, and cost, with radically contrasting applications, and care should be taken not to conflate these two atmospheric platforms. That said, military drone crashes shed some light on conservation drone crashes. The significance of a crash of a US military Predator MQ-1B in Djibouti emerges in complex interactions with its pilots and its monitored subjects. The wreckage, available for Djiboutians to see ruined on the ground, is a graphic materialization of covert military operations, a grounded example that even US air surveillance is subject to elementality and the failure of relative technicity or, as media studies scholar Lisa Parks writes, "the laws of gravity, software glitches, and bad weather" (2017, 151). The crash ensnares Americans and Djiboutians in "drone crash lore" (Crandall 2011, 12). Dialogue, journalism, and rumors proliferate in the crash's wake. Antidrone activism and calls for military transparency foment. Drone entrapments— asymmetrically concomitant relata within drone phenomena—are diffracted at and beyond the crash site. Crashes have consequences for terns and orcas, as for Djiboutians and Mexicans.

Another crash in 2011, this time of a small Mexican surveillance drone in El Paso, Texas, unearthed an array of converging and dispersing relata. The crash "opened the rituals of neighbors, the connectivities of machines, the routines of public agents, and the chorus of desert cicadas" (Crandall 2011, 2). Both centrifugal and centripetal forces were unleashed by the crash. "The drone crash . . . provides an exception, but also an amplification," writes media theorist Jordan Crandall, "a consolidation . . . and an agential dispersal" (2011, 13, 14). Endangered species have agencies incommensurable with Americans surveilled by Mexican drones, but crashes bare a diverse array of intra-acting relata in both contexts. Crashes force legal reforms for a more optimal future. In the aftermath of these crashes, we must "consider what might be salvaged from the wreckage" (Redrobe 2010, 22). Drones caused some wildlife wreckage in Bolsa Chica and the Puget Sound. Orcas and terns were damaged. Amid these impairments, wildlife wardens Molsberry and Mullins struggle to implement the legalities of blue governmentality. Flourishing awaits.

The "world-disclosing properties of breakdown" (Jackson 2014, 230) shift focus away from invention, innovation, and novelty and toward the forces of refuse, salvage, salvation, apology, and repair. As entropy—the eventual demise of complex to simple forms—and the contingencies of atmospheric exploitation erode stability, a role emerges for maintenance. "Broken world thinking" (Jackson 2014, 221) provides an ethical framework for approaching the crash's diffractions. Science and technology scholar Steven Jackson's (2014, 232) "ethics of repair" asks us to commit to caring for a world falling apart. Repairing drones, rehabilitating marine habitats, stopping poaching, catching criminals, and reforming regulations—this is the care of broken-world thinking.

The crashed drone and the threatened species—this is what remains for reworlding through multispecies care. Reworlding through repair is a form of multispecies justice that requires "nurturing and inventing enduring multispecies—human and nonhuman—kindreds" (Haraway 2018, 102). For humans, inventing kin through ecological repair may mean monitoring, mapping, and managing endangered species and their habitats, aided by digital, optical, mobile, and atmospheric technologies. It may mean surveillance, policing, pursuit, and punishment of human offenders. Care of endangered species aided by drones may not stop extinction, but it reveals how humans, technologies, and species are increasingly interdependent (Van Dooren 2014). Drones are a tool for technological reworlding that brings nature and culture into proximal but differentiated kinship.

What worlds do we make with conservation drones? These atmospheric remote sensors orchestrate a concrescence of elements, social spaces, electrons, photons, and species—as well as human institutions such as police departments, courts, and state legislatures. Some may argue that conservation drones are tools for "programming Earth" and "immediate environmental management" (Gabrys 2016, 4, 8). Or conservation drones can be interpreted as "vertical mediations" (Parks 2017), "exchangers between earthly processes, modified electric cosmos, human and nonhuman individuals" (Gabrys 2016, 13). Crashes diffract relata whose qualities exist between environmental control and out-of-control extinction.

Conservation drones are apparatuses for multispecies reworlding that hover between humans and endangered species, care and control, and the programmability and contingencies of blue governmentality. Drone crashes, by instantaneously and temporarily loosening otherwise knotted kin, refract the dualities of human pilot/endangered species and computer programming/environmental management.

The threat of a crash can also diffract multispecies thriving and cast light on the compromises inherent in wildlife governmentality. As we discovered in Bolsa Chica, some species of shorebirds are terrified by drones. The elegant tern's evasive reaction to the drone is instinctual. This is the animal's existential agency, its drive to survive. Understanding this requires more nuanced observations than mistakenly crashing a drone into a rookery. Studies have analyzed the effects of drones on seabirds. Unsurprisingly, drone disruption can be mitigated by higher flying (McEvoy, Hall, and McDonald 2016). Thus, there is a spatial threshold for intimacy between birds and machines. An early stage of building protocols of blue governmentality involves watching and responding to what is witnessed. From here, it is possible to incorporate birds' tendencies to fly off into plans for their flourishing.

Consider a nonmarine example, animal ethnographer Thom van Dooren's fieldwork with Hardshell Labs, a company that develops drones to frighten ravens (*Corvus corax*) from their endangered prey, desert tortoises (*Gopherus agassizii*), in the Mojave Desert in California. Their motto is "humane avian damage control" (Hardshell Labs, n.d.). Unlike conservation geographer Chris Sandbrook's (2015) human-centric critique of drone conservation, Van Dooren's approach is bird-centric. Drone pilots identify ravens in tortoise territory and fly toward and scare off the birds, causing them some stress as they ascend to drone-free skies. Their strategy builds on what ornithologists know about raven behavior and treats them as intelligent and cautious birds. And yet it is not innocent. It is a technologically enhanced management of tortoise and raven biopower. It is an example of technologies assisting multispecies thriving—a good-faith attempt to respond to animal consciousness and survival.

Hardshell Labs personifies the complex, compromised ethics of multispecies flourishing. Van Dooren reads their craft as a diplomatic solution to a problem that could be deadly, both for the ravens, whose eating of rare reptiles humans have deemed unacceptable, and for the tortoises themselves, who cannot afford to be eaten by corvids. Diplomacy is generative and exploratory. It begins with witnessing. From there it does not seek to control. Rather, the drone's flight scatters the ravens, nudging their behavior (Van Dooren 2019, 157). Leveraging the knowledge gained through exercising the patient "arts of attentiveness" (Van Dooren, Kirksey, and Münster 2016, 1), the scientists know that the drone will frighten the ravens, helping them achieve their objective of tortoise survival.

Diplomacy informed by attentiveness fudges control, and the term loses some of its purity. The "line between diplomacy and biopolitics (with its own forms of domination and control) is a blurry one. . . . Diplomacy and biopolitics are not mutually exclusive possibilities" (Van Dooren 2019, 160). The drone conservationists in this book begin in the spirit of attentiveness and diplomacy, respectfully working with animal intelligence in acts of existential flourishing. They are not averse to governance. Drone intervention may include disturbing nonhumans. Their conservation efforts blur control with biopolitical nudging. In storying the lives of humans and nonhumans in existential competition and collaboration, I witness not the hubris of conservation militarization but the humility of experimental conservation.

Van Dooren (2014, 2019) invites empathic realism for bird species on the brink of extinction, building their *Umwelt* into practices of resistance and mourning. One of his mentors, anthropologist Deborah Bird Rose (2011), weaves a tale of interdependence and colonial histories shared between the dingo (*Canis lupus dingo*) and Indigenous Australians. These multispecies studies offer narratives of animality that decenter speciesism, showing how nonhuman and human vitality move and shift each other. In emphasizing animal agency, multispecies studies might overlook the role of technologies that mediate animal encounters. In blue governance, however, the drone is an extension of relative technicity in the direction of ocean animals. It makes possible resistance to extinction and realizes an ethics of care. Whales, sharks, coral, seals, terns, and other ocean beings are not passive recipients of well-meaning human conservation. Their bodies, blood, breath, eggs, and images participate in a conservation phenomenon.

In his ethnography of the revitalization of Native American whaling in Washington State, anthropologist Les Beldo (2019) asks us to consider the whale's personhood. When the female gray whale indeed approached the small traditional whaling vessel, the Makah claimed it was giving itself up to support the people, while the antiwhaling activists thought that it was deceived, thinking it would receive the scratches on its callosities which tourists provide. Beldo (2019) summons us to decenter anthropocentrism by empathizing with whales. With anthropology's movement toward multispecies ethnography, it has begun to adopt what Lakota scholar Vine Deloria Jr. (2001) called "American Indian Metaphysics." Mohawk and Anishinaabe tribal member Vanessa Watts might be pointing to this metaphysics when she claims that many Indigenous people consider nonhumans such as trees to be contributors to a multispecies society that forms through interspecies communication. To not communicate is to not care, threatening indigeneity and its multispecies relations (Watts 2013, 23; TallBear 2011).

Blue governmentality is configured and disfigured by this animality: animal agency and instinct. Technologies and human institutions construct and deconstruct the blue governance of marine life. A proposal for animality asks us to approach marine mammals, fish, and birds with attentiveness and diplomacy. The goal of conservation is preserving animal life. Conservation drone phenomena coalesce to improve the prospering of these animals. Forging convivial, caring, and diplomatic companionships with multiple species breaches anthropocentrism. The control of blue governmentality is not without annoyance, pain, and death. It requires sacrifice and the intelligent mobilization of animal agency and impulse for thriving.

ENTROPY

This chapter has investigated the entanglements of disintegrating ecologies, tumbling drones, and intervening humans. It provides an account of drones' enmeshments with terrified terns and starving orcas on the West Coast of the United States. In the exact moment of a crash, there is no conservation drone without its intra-action with endangered species. Connected to the drone is a human pilot, linked by an exploitation of the electromagnetic spectrum. Suspending the drone is the lift provided by the gas molecules of the atmosphere. The crash makes obvious the interlacing of drones, species, humans, the electromagnetic spectrum, the Earth—and countless others, as each of these agents is constituted by other smaller relata. With the hardware broken and the wiring exposed, a techno-mammalian-elemental phenomenon is revealed.

The crashing of a drone in an elegant tern nursery in California provoked the parent birds to abandon thousands of eggs. The culprit has yet to be captured. Likewise, the threat of drone crashes in the Puget Sound near orcas revealed the impacts of wildlife protection laws and their negotiation. These crashes, or their risk, reveal the slippages of wildlife regulation, blue governmentality, and ocean/cultures. Conservation laws are diffracted by the crashing drone. Law and regulation escape their drone phenomenon. Oceans are not cultured by human law because legal applications often fail.

Crashes occur—this is true for drones, species, and robots today and tomorrow. Conservation drones are apparatuses of multispecies and computational reworlding. This process, however, will never be final. It will be faulty and will emanate in a tightening morass of technologies, elements, and living beings, with people decreasingly the driving actants in the phenomenon. Like orcas and terns, humans too will be increasingly (but never totally) entrapped by technologies that are designed to prolong vivacity but are prone to failure. Inventor and complexity theorist James Lovelock (2019) predicts that smart robots will preserve the Earth for their own survival—even androids need stable Earth systems for their long-term survival. Thus, eco-programmability will increase but humans may not necessarily be the dominant actant in this future phenomenon. Automation and AI, climate crises, and other para-human factors are challenging and will increasingly challenge anthropocentric theories.

The drone crash is not merely a metaphor for the sixth extinction. It is not only a symbolic representation of the Anthropocene. Rather, the drone crash and the decline of marine biodiversity are of the same substance. Entropy, the turn to simplicity, links the two materially. Like a diverse ecosystem, the functioning drone realizes complexity. The failing drone and the dying world are in a state of decay. From the whole they become parts, and from parts, fragments. A homecoming to cold geology approaches.

Drones are constructed from mineral, electrical, and chemical processes. When they crash, they result in a mess of metal, propellers, plastics, plants, soil, and sand. Offspring are abandoned and feeding disrupted. Crashes challenge regulations and invite modifications to existing legalities. Blue governmentality—the exploitation of relative technicity, elementality, and interventions for the extension of nonhuman life—emerges broken, yet repairable, from this twisted mesh. Here, the drone hovers between death and its care, floating between entropy and its negation.

7

SHARK NETS AND INTIMACY

It is spring 2021, and rainstorms fill the bays around Sydney, New South Wales, Australia, with urban runoff. This evening, instead of the usual clarity, the water is opaque. It is thick with something. Chemical particles are threaded with microplastics. An update of the daily pollution forecast, courtesy of the New South Wales Office of Environment and Heritage (OEH), informs swimmers whether pollution is unlikely ("enjoy your swim!"), possible ("take care"), or likely ("avoid swimming today"). Whether it is an invitation or a warning depends on whether it has rained and how hard. I assume there are sensors at the beaches that perceive the presence or absence of algae blooms, fecal coliform, transmission oil, and tire lubricant. After a week of showers, the advice is to "take care" this dusk hour in Clovelly Bay, a five-minute walk from my apartment. Despite the warning, I don a short spring wetsuit and a mask and jump in. As I swim out from the shore in this narrow bay, its sand, dusky flatheads (*Platycephalus fuscus*), and stingarees (*Trygonoptera testacea*) give way to sandstone shelves, urchins (*Echinoidea*), gobies (*Eleotridae*), and sea slugs (*Nudibranchia*). "Bluey," a fat resident blue grouper (*Achoerodus*), swiveled its left eye as I passed, wondering if I might do as I have often done before and flip over rocks it cannot, revealing dinner crustaceans. I relent and expose a few morsels for the fish, the stone thudding in the dense water acoustics. Swimming out and over the sandstone rubble that guards the bay's mouth, defending the sheltered cove from the rough southern swell, I am quickly in what feels like open ocean.

Visibility improves in this cleansing current as the sun sets behind the beach, dampening the coast in ocher twilight. Kelp and blackfish sway with a set of waves that roll over my head. In deeper water I enjoy a free dive, charging my lungs, flipping my legs over my head, and allowing my weight to work my body vertically down. A few kicks and I grab a bottom-dwelling cunjevoi (*Pyura stolonifera*), a tough sea squirt that was filter feeding (unadvised on this day, according to the OEH), and rest. Coming to the surface, something is different. I have swum and snorkeled here numerous times and know the topography well—here is where I saw the giant cuttlefish (*Sepia apama*); over there the squid (*calamarius*) inked me; and here the submerged headlands drop into an abyss. But there was a new formation. Perhaps in the dying purple light I had drifted farther than where I had assumed, and this novelty was a geoformation, a rock wall I forgot. But this was not geological. It was zoological. It was moving, color-shifting—a shimmering golden torpedo body that propelled itself from an undisclosed power source, its weight, trajectory, and forward momentum carrying it like an aircraft carrier. From its hydrodynamic head to its thick body and sculpted tail—it did not stop. Three meters long, as wide as three of me, its tail as tall as I (or so it felt), and more elegant in its arabesque curls, it cruised by. It was a copper shark (*Carcharhinus brachyurus*), known locally as a bronze whaler and reviled by spear fisher folk for ripping catches from their buoys. It eventually passed. I was split between wanting to see and know more, to follow it for a while, marauding the dark crevices with it—or to get the hell out of the water. My frightened amygdala beat my curious prefrontal cortex in the cranial debate, and I let this shark continue its evening purview of the seafood buffet. A wave of adrenaline shot from my head to my feet. Then a second and a third jolt of survival biochemistry hit, and I was back onshore, haunted and blessed by the encounter. Brief and profound, such underwater encounters offer moments of multispecies intimacy. This was the shark's element—this area is known as Shark Point.

FIGURE 7.1. Original shark net at Coogee Beach, 1922. From the collections of the State Library of New South Wales [a6415019/ON 30/box 55 no. 675] (Mitchell Library). Source: Foster 1922.

This diurnal hour was feeding time, and the post-rain visibility offered sharks the cover needed for sudden, fatal, sustaining strikes. But not on me, not this night—and almost certainly not ever.

Shark bites are exceedingly rare. The first record of a European being bitten by a shark in Australia happened in 1791 (Gibbs et al. 2020, 192). With daytime public bathing outlawed in much of Sydney from 1833 until 1903, fatalities and bites remained low—only three injuries and six deaths happened between 1790 and the 1870s (Gibbs et al. 2020, 192). Throughout the world, however, in Australia, the United States, South Africa, Reunion Island, and the Bahamas, shark bites are on the rise (Tate et al. 2021; McPhee 2014). Several explanations are offered: more humans, more human bathers, and more humans accessing formerly isolated beaches (West 2011, 744). Worldwide, sharks kill about ten humans a year (Tate et al. 2021). On average in Australia, 1.1 people die annually from shark bites, with another 5.9 sustaining injuries (West 2011). Compared with 136 coastal drownings in Australia in 2020 (Surf Life Saving Australia 2021), 1.1 annual deaths is a small number indeed.

Near the waters I swam in that evening, it had been one hundred years since eighteen-year-old surfer Milton Coughlan and twenty-one-year-old

FIGURE 7.2. Coogee Pier and shark net, circa 1930. Source: Foster 1922.

Mervyn Gannon lost their lives to bites in neighboring Coogee Beach in 1922—the last such fatal interaction. These tragedies provoked several responses in the days that followed: people refused to swim; shark bounty hunters were called in to catch and display slain sharks; thousands gathered to watch sailors from New Caledonia dive with knives in their teeth to hand spear the sharks; and someone suggested that the bay be bombed from the air to rid it of sharks (Weeks 1992). That year, Randwick Council built a shark net from headland to headland, across much of the beach (figure 7.1). But before its dramatic public opening, a storm surge destroyed it. The council continued to build shark barriers. In 1929 a major pier was built into Coogee Bay, and using a pylon as one anchor and the headland as another, they strung a second shark net (figure 7.2). Entry into this "safety-first surfing" area cost one penny. But once again, heavy seas hammered Coogee, demolishing the pier in 1934. The shark net was kept up until the 1940s, when World War II and a metal shortage made repair impossible. But following the recommendation of the New South Wales Government's Shark Menace Advisory Committee, by 1937 many beaches had been netted (Pepin-Neff 2012).

Fish nets are ancient, quintessential bodily prostheses. Modern *Homo sapiens* were a sparsely populated coastal species early in our history, catching and cooking sea life in South Africa around 100,000 years ago (Marean 2012; Henshilwood et al. 2001). Later, around 42,000 years ago, humans developed deep-sea fishing and, with it, a taste for tuna (*Thunnus*), as evidenced at the Jerimalai shelter in Timor-Leste (O'Connor, Rintaro, and Clarkson 2011). These folk used primeval marine traps such as nets and hooks to extend the grasp of their hands into and below the ocean. Traps, according to anthropologists Alberto Corsín Jiménez and

Chloe Nahum-Claudel, are "autonomous, animate and deadly technologies." As "proto computers," the "first robots," and "primitive information processors," traps are technologies that network animals and humans (Jiménez and Nahum-Claudel 2019, 397). They store and relay energy with brutal efficiency. Ideal prostheses for overcoming time and space and executing force from afar, traps work temporally and territorially. Nets bring together the human and the nonhuman, the living and the dead, in an anticipatory cybernetic embrace. Situated for the future, traps are existential technologies of intimacy, brutality, and mortality.

The shark net program continues to this day. The New South Wales Shark Management Strategy constitutes the world's longest-running lethal shark management in the world (Gibbs et al. 2020). Fifty-one nets are placed near New South Wales beaches. North of us in Queensland, eighty-five beaches have nets or drumlines—baited hooks floated by buoys and anchored in place. New Zealand, Egypt, Russia, Mexico, and the Seychelles have also strung shark nets or drumlines. Many beachgoers I have spoken with think the nets prohibit sharks from coming near the beach. But this is not true, nor the objective. The nets cover only a fraction of each beach. Set adjacent to the shore, suspended by buoys, and anchored to the bottom, in New South Wales the nets are 150 meters long and 6 meters deep. In Queensland, the nets are between 124 and 186 meters long and 6 meters deep (figure 7.3). Sharks can swim under, around, and, if they are enterprising, over the net.

Contrast that length with that of two Sydney beaches, Coogee and Bondi, which are 400 meters and 1 kilometer long, respectively, and their waters are significantly deeper than 6 meters. These nets are not barriers but gill nets for sharks—and whatever else they catch. They are fishing technologies that entangle more than fish. Nets, as theorized by geographer Catherine Phillips (2017), capture sharks, people, and other organisms in intimate and existential embraces. Situated as they are in a sea that magnifies entanglements through its fluid and mixing qualities, nets both empower and complicate terrestrial, human political power (Lehman 2013). Evading such nets, as countless sharks do every day, exposes the "multiple individual micro-situations in a variety of environments where animals may respond unpredictably, resist human power, and even exercise power themselves; but these micro-situations are 'invested, colonised, utilised, involuted . . . by ever more general mechanisms and by forms of global domination,'" as Foucault wrote (quoted by Palmer 2016, 121).

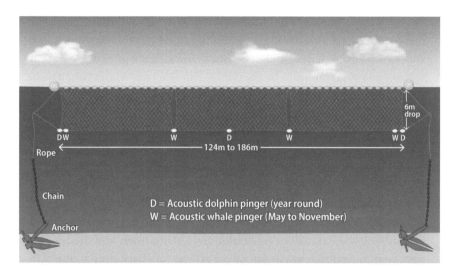

FIGURE 7.3. Diagram of a shark net. Source: Queensland Department of Agriculture and Fisheries 2021a.

The aim of the Shark Control Program, in the calculation of the state government, is to "reduce the chance of people being killed or seriously injured by sharks in Queensland. This is achieved by removing large dangerous sharks from 85 popular beaches along the Queensland coastline through the use of nets and baited drumlines" (Queensland Department of Agriculture and Fisheries 2019). The program targets nineteen shark species—including several species protected by the Environmental Protection and Biodiversity Conservation Act, such as the white shark (*Carcharodon carcharias*) and the shortfin mako shark (*Isurus oxyrinchus*)—a targeting criticized by conservationists (Vincenot and Petit 2016; Queensland Department of Agriculture and Fisheries 2019). In its history, the Queensland Shark Control Program has caught 50,000 sharks (Roff et al. 2018). Specifically, between 2001 and 2019, 8,692 sharks, including 2,504 tiger sharks, 1,612 bull sharks, and 70 white sharks were snared. Many nontarget fish and animals have also died in New South Wales as bycatch, including twenty-three species of sharks and fourteen other organisms including dolphins, whales, seals, penguins, and finfish (Green, Ganassin, and Reid 2009). Fifty-six humpback whales have been caught in Queensland (Bolin et al. 2020). Despite this cull, in sixty years, shark bites persist, with thirty shark bites—including two deaths—in Queensland beaches that are fitted with drumlines or nets (Australian

Shark Incident Database n.d.). Similarly, in eighty years, there have been thirty-four shark interactions—including one fatality—on New South Wales beaches with nets and drumlines. The efficacy of this program is suspect and under inquiry by the public, shark activists, and the government. One thing is clear about the nets and hooks: they are effective as indiscriminate fishing tackle.

Throughout Australia and other localities around the world, drones, often equipped with computer vision and artificial intelligence, are being trialed to identify sharks and dangerous swimming conditions, and to alert swimmers. In some cases, these lifesaving drones can drop flotation devices and other life-support equipment. If successful, funded, and supported by the public, such programs can replace nets and drumlines, minimizing the harm done to sharks. Nets, drumlines, and drones present clashing approaches to the co-occurrence of sharks and humans in coastal waters. The nets and drumlines segregate space by species; drones open the nautical borders to interaction between sharks and humans. Both make possible contrary modes of intimacy: one with dead and dying sharks, another with their living kin.

Shark populations are on the decline worldwide. The numbers vary, but around 100 million sharks are caught per year (Worm et al. 2013). Based on a study of the records of catches of the shark-netting program between 1950 and 2018, it is clear that shark populations are in free fall in Australia (Gibbs et al. 2020, 195). Between 74 and 92 percent of hammerheads (*Sphyrnidae*), whalers (*Carcharhinidae*), tiger sharks (*Galeocerdo cuvier*), and white sharks (*Carcharodon carcharias*) disappeared in this period (Roff et al. 2018). The cause may be the nets, the targeting of sharks by fishing industries, climate chaos, or something else.

To protect sharks, citizens across Australia protested nets and drumlines. Thousands of participants in the No WA Shark Cull campaign gathered at Perth's Cottlesoe Beach in Western Australia in 2014 to reject the state's shark cull (Donnison 2014), and protesters have gathered at numerous other beaches throughout Australia (figure 7.4). Activists tracked the Department of Fisheries vessels while they cleared the drumlines of catch and documented the brutality of the disposal and release of nontarget or undersize sharks. In one instance, activists showed how the contractors, who are paid $5,700 a day for this work, cut the jaw of a young tiger shark to free it from the hook (Gibbs and Warren 2014, 101; Luke 2014). The beach protests, direct actions that remove bait from the drumlines, rendering them useless, and gruesome footage circulating on

FIGURE 7.4. Drone shot of beach protest Nets Out Now, Coolangatta, Queensland, August 2021, after a humpback whale was freed after forty-eight hours of attempted rescue, only to leave still dragging ten meters of chain and net. Source: Andre Borell.

social media may have helped convince the government of the state of Western Australia to abandon their shark cull in 2014 (Weber 2014).

Many activists and citizens hope for a similar outcome in eastern Australia. Taking a page from the successful methods in Western Australia, activists in New South Wales and Queensland have begun to monitor the shark nets and drumlines along our coast—and the subcontracted fisherfolk tasked with checking them. One of the key figures in this methodology is schoolteacher and activist Jonathan Clark, who, beginning in 2015, drove one to two hours with a team from Brisbane, Queensland, to the Gold Coast or the Sunshine Coast every other weekend to check the shark nets and drumlines. Clark and his team of three would leave Brisbane around 4 a.m. Once on the sea and near the nets, they would jump in the water, swim alongside, and look for caught fish and animals. Clark, with stringy hair and a rough salt-and-pepper beard, explained to me the early days of developing this protocol:

> Adam, we didn't really know what we were doing, just testing the waters. I can tell you just quietly that my first-ever snorkeling expe-

rience, the first time I ever put a bloody snorkel and flippers on, was pulling up at a drumline, jumping off the back of a jet ski, sticking my head in the water, getting filled up with bloody exhaust fumes from the jet ski, looking in the water, talking to Pauli, who's driving the jet ski, and asking, "What am I looking for?" [He responded,] "Follow your eyes down the rope and see if . . ." And then the moment of realization occurred to me that I'm hanging over a shark hook looking to see if there's a shark on it. (personal communication with the author, March 11, 2021)

As ocean activists, they did not want to find any dead or dying turtles, rays, sharks, birds, dolphins, or other species. But capturing and distributing images of drowning sharks, sea mammals, birds, and reptiles twisted into nets just outside of popular beaches would help galvanize the public against these mortal shark-mitigation programs. So they dove, again and again, monitoring the eleven shark nets and thirty-five drumlines in their region with no shade, in swell, chop, and salty wind—for years.

Since those formative days, their technics and technologies have improved. Today, their mouths are no longer filled with exhaust, as they acquired a larger boat and learned to back roll into the water. As they approach the shark net, someone sits up on the bow and another person sits on the stern. Wearing polarized glasses, they look down into the water to see if they can see anything. The crew usually includes a free diver who is skilled at underwater photography. Coordinating relative technicity with turbulent elementality is difficult, as Clark explains: "it's one thing taking photographs on the surface, but multiply that by bloody ten to do it underwater, especially in a stressful situation on a shark net or a drumline" (Clark personal communication 2021).

The activists have tried to orient their vision along the vertical axis of the shark net with various degrees of success. The underwater drones they deployed did not have enough propulsion to navigate the powerful currents in eastern Australia, and action cameras like GoPro on monopods are difficult to deploy because the cameraperson cannot see what they are doing without a live video feed. Today, when they approach a drumline or net, they lean over the side of their vessel and use a bathyscope, a kind of reverse periscope, to magnify their eyes and see if any species are hooked.

If there are, they jump into action with specialized procedures depend-

FIGURE 7.5. Holly Richmond and a dying tiger shark caught on a drum line, Gold Coast, Australia. Source: Borell 2021.

ing on whether the caught organism is a shark or not, alive or dead. They back roll into the water with high-definition cameras like the Canon 5D Mark IV, which is secured in a waterproof housing. Then they float for an hour, noting conditions, collecting images, making phone calls, and constantly repositioning the boat to stay near the net, divers, and dying marine life.

This survey is not just for spectacle. They conduct a numerical assessment of the efficacy of the shark-mitigation scheme. Any nonsharks found demand an immediate phone call to the Queensland's Shark Control Program and the Marine Animal Release Team, asking, as Clark relayed to me, "'When are you coming out? We've got a [sting]ray on the line. It's a nontarget animal.' Or 'We've got a turtle on the drumline. . . . When are you coming out?' Sometimes the response is: 'Oh, it's only a ray, Jonathan. We'll just pick that up next time a contractor is out'" (Clark personal communication 2021). It is illegal to do so, and therefore, Clark and his colleagues will not tell me whether they free the fish, reptiles, or marine mammals they find clinging to sentience yet caught in the nets. But as I describe below, some do, producing multispecies intimacy through an act of disentanglement.

Marine biologist Holly Richmond is often on these reconnaissance missions (figure 7.5). She dives down a drumline, recording sharks living and half-living, and is deeply moved by these encounters, saying,

> My first time swimming with a tiger shark was on a drumline and she was literally taking her last breaths in front of us. She had been there since the morning, and she was suffocating on this drum-

line, and I was able to get really close to this animal. Looking at the details of this animal was insane, the patterns on their skin. And people always refer to sharks' eyes to be lifeless and soulless and dark and black. But looking at her eyes, they are light-colored brown, and they were so beautiful and deep. You could be looking into the eyes of a turtle or your own dog at home. That moment really made me feel connected to these animals and made me realize that they're just crying for help, and we're just endlessly killing them out there. (Borell 2021)

Shark activist Madison Stewart had a similar experience diving a drumline. She confessed,

I got in the water thinking [the shark] was dead but then its eyes were still moving, so it was just this animal that I'd always associated with such power and presence in the ocean, and it was just lying there dead. It was like going into your yard and seeing your own pet dog hooked up on a drumline. Most people talk about *getting in* the water with sharks for the first time and their hands are sweaty and their heart's racing and they're freaking out. That is exactly how I felt *getting out* of the water after filming a dead one. . . . And the saddest thing about all of it is living in a country where people kind of rooted that on, and wanted that to happen, and in their eyes, that [catching sharks] was a good thing. (Borell 2021)

Clark, who regularly edits the footage collected by divers like Richmond and Madison, said, "when I look at that footage afterwards . . . when I listen to a diver crying underwater, when I listen to a diver spontaneously apologize to the animals under there, it gets me, when I am sitting there in my living room, watching it on a computer to try to get that message out. How do we do that effectively? And how do I look after my crew when we're doing that? And we are not just doing it once. We are doing it again and again and again" (Borell 2021).

A materialization of the theories of entanglement (Barad 2007), the nets congeal the physical and affective linkages between humans—our exosomatic technologies, the filtering effects of fluid elementalities, and the existential livelihoods of marine life. For these activists, confronting shark mortality with their submerged bodies is an essential, initial step toward eradicating the nets. Here are two bodies—one human, another shark; one cultural, another oceanic—that are sharing an exposure to harm and willingness to survive. Sharing oceanic elementality, these activists join in a multispecies "fellowship with suspended and suspensible others" that results in a felt experience of "susceptibility and embeddedness" (Choy and Zee 2015, 217). Feminist ecocritic Stacy Alaimo might describe these two bodies by referencing not their objective individuality, but rather their subjective relations as transcorporeal beings rapt in ego-dissolving seas. This shark/human connection "floats in a productive state of suspension, between terrestrial human habitats and distant benthic and pelagic realms, between aesthetic estrangement of sea creatures and the recognition of evolutionary kinship, between mediated, situated, and emergent knowledges and an ethico-aesthetic stance of wonder" (Alaimo 2012. 490). In these transcorporeal connections, the wonder felt toward sharks shifts into mediated activism. Negative thanatopolitical intimacy is leveraged into affirmative zoëbiopolitical intimacy. This means coexisting with sharks, sharing their sea, and accepting the fear and the 1.1 annual fatalities that this may mean for Australians. Mediating these dangerous, deathly encounters spreads their affect to audiences, ideally galvanizing them to act against the fatal shark control programs.

SHARK DRONES

Since they were put in place in Queensland in 1962, the shark nets have moderated a lethal relationship between sharks and humans. The nets entangle more than sharks in intimate, existential unfoldings. Drones and whales also get caught in the mess. To explore the issues of bycatch and drone activism, I spoke with activist, drone pilot, and filmmaker Andre Borell, who produced the 2021 film *Envoy: Shark Cull*, and who advocates for the retirement of the lethal shark program and its replacement by surveillance drones capable of seeing sharks and warning swimmers. I wanted to know more about one compelling drone shot in his film.

He had storyboarded a sequence in which a drone flies across a shark net along Surfer's Paradise, Queensland, and then tilts up to reveal the towering hotels, revealing the net's proximity to this iconic location. On the day of the shoot, the drone cinematographer called Borell, saying, "'You're not going to believe this but there's a whale in the net.' I went, 'No, no, no, you're kidding. There's no way that coincidence happened.

Like, the day we're shooting that shot, there's a whale in the net?' He replied, 'No, no, seriously. There's a whale in the net. Get down here.'" Borell jumped in his vehicle and raced to the beach, notifying rescue teams and journalists on the way—"Because you always want these events to be seen, not just swept under the rug" (personal communication with author, October 28, 2021). He joined the drone pilot on the beach, and they continued to fly above the trapped adolescent humpback whale, waiting for the Queensland Marine Animal Release Team to arrive and cut the whale free.

Twisted in the net, the juvenile whale appeared exhausted, its breathing labored as it dipped with the waves. The drone too was running low on energy, its battery quickly fading. Borell called other drone operators on the Gold Coast, compelling them to immediately come to the shoreline with additional drones and extra batteries so that they could keep a drone constantly aloft. With a drone in the air, Borell ran to the local surf shop and tried to rent stand-up paddleboards so that they could head out to the whale with a scuba diver's knife and try to cut it free. They knew that they would be subjected to substantial fines for approaching the whale and the net, "but it was a risk we decided that we were probably going to have to wear. Because if it came down to this whale surviving or not, so be it" (Borell personal communication 2021).

Gearing up to depart from the beach, they took a final look through the drone and noticed a "tinny"—a small aluminum fishing boat—cruising directly toward the trapped whale. The captain pulled up right next to it, and before they knew what had happened, he had jumped in the water, his abandoned boat floating away. Sixty seconds later, the whale was free. The man triumphantly punched the sky as the whale squeezed out of the trap. The drone operators waved Borell off. Interviewed onshore by a local television network, this man, who went by the single name Django, said, "I'm a typical Aussie male, I do stuff first and think about it later. There was no real thinking, I saw it and that was it, you just get going and get in the water" (Cooper 2020) (figure 7.6).

An hour passed, and the Queensland Marine Animal Release Team finally turned up, replaced the shark net damaged by the whale, chased down Django, and issued him a citation. Django was served with two Fisheries Infringement Notices, one for entering the exclusion zone and another for interfering with the shark net. Based on the publicity from the drone cinematography, a crowdfunding campaign raised money to pay the penalty. Eventually, the government backtracked out of the fine and

FIGURE 7.6. Drone shot of Django diving in the ocean to free the young humpback whale from the shark net, Gold Coast, Australia, 2020. Source: Andre Borell.

let him off with a warning. According to Django, shark nets "are a waste of time . . . sharks just swim around them" (Cansdale and Cummings 2020). He donated the crowdfunds to the local Sea Shepherd branch.

In this tale, a drone flies above a young humpback whale that is entangled in a net designed to catch sharks that might pose a threat to humans. Django saw the whale blow from four hundred meters away and came—not because he saw the drone hovering above, but because he saw the whale blow near the net. Nevertheless, the drone's footage, routed to local television news, social media, and a feature-length documentary, transmitted the plight and brutality of bycatch. The drone could not see through the opaque sea surface, but it could provide an atmospheric account to pair with the evidence of the indiscriminate bycatch associated with shark nets. This is an outcome of drone technicity—its mobility to quickly fly from shore, share airspace with other drones, and depict in high resolution the effects of shark nets on whales. What the drone footage does document—as do the television interviews with Django, accompanied as they are by the drone footage of him abandoning his vessel and jumping in the water—is not the dead and dying multispecies intimacy of Richmond, Stewart, and Clark, but of a struggling yet living whale. This drone mediation of living organisms exists in a world in which nets are slashed, and marine species and humans swim free. As I discuss below, if funded indefinitely, the trials to use drones to identify sharks and warn swimmers, currently underway in Queensland and New South Wales, might make this free movement and coexistence possible.

Heroic acts such as Django's rescue and Borell's drone videography provoke public outcry and influence the government to research non-lethal alternatives to shark nets and drumlines. One example is the Cardno Report, prepared for the Queensland Department of Agriculture and Fisheries (McPhee et al. 2019). It recommends the use of nonlethal shark barriers and deterrents for northern Queensland beaches—such as Cairns and Townsville—where oceanic visibility is low and wave energy weak. These are good conditions for underwater barriers, but poor for drone vision from above. Conversely, it suggests the application of drone surveillance for the southern Queensland beaches—locations like the Gold and Sunshine coasts—where the ocean is clearer, but wave energy is too strong for shark barriers (McPhee et al. 2019, ii). Elementality and relative technicity conspiring for transformative conservation.

When Clark went up to Cairns, in northern Queensland, to attempt to monitor the shark nets, the lack of visibility was evident. They pulled up beside the drumlines, put GoPro cameras in the water, and could not see fifty centimeters. The water was too opaque for action cameras or drones. Another technology was needed, such as the Eco Shark Barrier, which mimics a kelp forest and keeps sharks at bay (but not crocodiles, another problematic species for human bathers in northeastern Australia). Bubble curtains, which release a wall of percolating gas; electricity-leaking cables that irritate shark senses, such as their ampullae of Lorenzini; electrosensors; sonar; drones; and SMART (Shark Management Alert in Real Time) drumlines are all elemental options, each providing a relative technicity for coexistence. These and other nonlethal applications are techno-elemental: electromagnetic fields and gas, and auditory and chemical stimuli that permeate the sea, deterring shark activity. These technologies exploit the elements to dissuade shark intimacy.

Nonlethal shark mitigation is a form of blue governmentality that must first address the relationship between relative technicity and elementality. The Cardno Report outlines how extensions of human intentionality toward sharks through the atmosphere and within a sea of varied opacity and wave action might care for and control shark agency. The report encourages Queensland to embrace what it says is a "philosophy of coexistence" with sharks (McPhee et al. 2019, 1). The strategy must adapt technologies to the elements and employ localized approaches when necessary, depending upon the force and clarity of the sea. In the future, swimmers in the north of Queensland may swim behind a permeable wall of bubbles, electromagnetic radiation, or kelp-imitating rubber fronds bolted to the seafloor. In the south of Queensland, drones may enable greater interaction. Shark and human territoriality will be less clearly but also less lethally differentiated. Nets, baited hooks, bubble barriers, fake seaweed, electrical boundaries, or drones—singular technicities adapted for elemental peculiarities and animal agencies make divergent embodied intimacies possible.

Mobilized by public opinion and informed by scientific evidence such as the Cardno Report, today both New South Wales and Queensland are trialing drones for shark mitigation. The SharkSmart Drone Trial featured drones at beaches in southern Queensland flying during weekends, public holidays, and school holidays from September 2020 to March 2021. The trial aimed to detect sharks, warn swimmers of potentially dangerous sharks, document specific species and their behavior, and teach and test artificial intelligence to identify sharks. Flown by volunteers from Surf Life Saving Queensland who had acquired a remotely piloted aircraft operator's certificate, drones were in operation from morning to midday—afternoons were excluded because of the likelihood of gusty winds. When a shark was spotted, the drone flew lower so that it could follow it, attend to its behavior, and measure its size. Lifeguards assessed the situation and sounded an alarm to evacuate the waters when risk was elevated. In 1,565 flights covering 626 kilometers, the trial spotted eighty-seven sharks, including fourteen over two meters in length (figure 7.7). Twice they signaled the evacuation alarm. According to nearly two thousand public responses, 96 percent of the public supported the program (Queensland Department of Agriculture and Fisheries 2021c). No human or shark was injured (Queensland Department of Agriculture and Fisheries 2021b). The trial of the drone shark program was a success, and the governments appear committed to funding the program.

The benefits of drones for shark mitigation are many: they are non-lethal, more affordable than helicopters, offer real-time monitoring of beaches for other hazards, and are supported by the community. The optics on shark-detecting drones can feature hyperspectral sensors and polarizing lenses that enhance underwater seeing. Fluid lensing software can correct the distortions of light as it passes through water. Artificial intelligence can be taught to identify sharks. Shark drones can transtechnologically integrate with other platforms. SharkEye (https://sharkeye.org),

FIGURE 7.7. Drone shot of drone follow-ing shark in Austra-lia. Source: Andrew Colefax.

in California, and Dorsal (https://www.dorsalwatch.com), in Australia, for instance, offer phone-based shark alert systems informed by drones and artificial intelligence. This artificial intelligence will allow drones to "'see' further below the water's surface than human observers in manned aircraft" (Butcher et al. 2019, 703). Drone vision is less elementally dis-rupted than a human peering out of a helicopter. For example, "strong winds and rough seas may actually impair marine fauna sighting rates less for drones than for manned aircraft, because drones are flying sub-stantially slower and lower" (Butcher et al. 2019, 708). Shark surveillance drones can perform other functions—such as dropping flotation devices to struggling swimmers. This drone shark surveillance may eventually be automated. The drones will take off, survey beaches, identify sharks, alert swimmers through speakers, land, self-charge, and repeat—all with little human oversight. Australian beachgoers have already con-fessed that they do not mind the privacy infraction that drones might pose in exchange for the safety services they provide (Gibbs and Warren 2014). Such a service reorients how space is conceived and segmented. First without nets and drumlines, and then without a human pilot, shark coexistence may proceed through omitting actants. Many negative bio-politics are no doubt possible. Things will go wrong—crashes, false posi-tive and false negative identifications, and so on—but this change is one that will potentially produce fewer shark deaths.

Change is possible because it has happened before. Beginning in 1989, the shark nets in New South Wales were taken out during the winter months to minimize the chance of catching a migrating humpback whale. The state government reduced the time between checking the nets from ninety-six to seventy-two hours, decreasing the duration animals are hooked and trapped, and increasing the chance of survival for nontarget species that would be released if found alive. They took the nets out around Cairns, near the biodiverse Great Barrier Reef Marine Park, in 2013. Cetacean acoustic alarms were installed on drumlines and nets. All these measures were enacted to decrease unnecessary death. The drone program is the latest relativization of technicity from necropolitics to the zoēpolitics of blue governmentality. In the process, how humans approach the spatiality of the ocean—as swimmers, technologists, and geographers—will also metamorphose.

TERRITORIALITY

Imagine if the nets were a real barrier that was impervious to sharks—a surface-to-seabed, headland-to-headland-wide net that permanently locked sharks out of a bay and only allowed the ingress of ocean water, small fish, seaweed, and plankton. Human and shark territories would be firmly marked "in" and "out." This is what the council envisioned for Coogee Beach back in 1922. That net did not work, but it materialized a sentiment that our two lifeways—sharks on the one hand, and humans on the other—cannot intersect, or they can do so only on human terms, and only when humans want to cross that barrier to make sharks into food, curiosities, or sport. In actuality, shark nets are permeable. Short and shallow, they are ineffective deterrents but decent killers. That is their purpose. They are gillnets for capturing and killing, not tools for dissuading interaction. The nets do not actually work to deter sharks from swimming to the shoreline, as some humans have claimed. Consequently, with the soaking of these partially effective nets comes the submerging of a social fact. The tacitly accepted but unspoken understanding is that we cannot live with sharks; nor can we ethically eliminate them.

With the originary technicity of normative bodies, modern humans traditionally lived along and gathered resources from the ocean. Deeper engagement with the element required technologies—snorkels, fins, scuba, Jet Skis, rods, reels, surfboards, rigid inflatable boats, fishing vessels, and so on. The partiality and brevity afforded by these technologies of marine visitation remind us that our territorial dominance is temporary. The ocean refuses permanent undersea or stable surface constructions. Ocean/cultures, not oceancultures, remain the rule.

But still, we try to execute our care and caring control through conservation of the sea. This move represents the evolutionary drive of originary technicity—traveling over, grasping, containing, and covering new things and spaces in a wager with and against the tumultuous elementalities of the sea and the air. Despite—and often because of—elemental disarray, we find intimacy in moments of intra-action and engagement: a masked great ape and a shark passing each other on an ill-advised evening swim; snorkeling activists witnessing a shark's final breath on a hook; a tinny carrying a diver that releases a tired and relieved adolescent whale to continue its northern migration; swimmers receiving the audio waves of a shark alarm and quickly departing a wave set to make way for a big foraging shark. Technicities mediate elements. They afford intimacies. Care and control follow. Coexistence is possible.

Just another five hundred meters out from my neighborhood beach are the fatal shark nets. Near the shore, close to swimmers, and at the whim of the tides and waves, the nets are at the coastal interface between the land and sea. As liminal membranes, the nets let flow the sea to the shore, and human values of the land into the sea. Beach formations, influenced as they are by submarine topography, lunar and solar gravity, and oceanic particulates—neither entirely land nor sea—are a challenge for humans to govern. Geospheres and hydrospheres offer their own nonhuman protocols for geopolitics, not the politics of international states, but of Earth systems. Shifting sands and wandering sandbars mean the sediment is not solid. This terrain refutes firm foundations, strict boundaries, and the robust legal definitions that depend upon stable, linear, hard, Cartesian ground. The limited functionality of blue governmentality depends upon going with but also directing the flow of this geopolitics and its meandering and mediating fluidities.

For shark geographer Leah Gibbs, the nets are an "old and outdated political technology" (2018, 209). As an antique method of "governing ocean space" (Gibbs 2018, 207), the net is an attempt to segregate the fluid ocean and differentiate human from shark territory. Faulty and permeable borders as they are, the nets fail to define an inside or an outside. Nevertheless, the nets strive to assert "territory through bounded space and through social power exerted by violence" (Gibbs 2018, 218). The Australian states misunderstand shark behavior and argue that they are territorial (they remain in specific locations) when they are in fact migratory. Marine geographers Christopher Bear and Sally Eden (2011) implore us to think not only oceanically, but with fish behavior. They argue that a better understanding of sharks' movements and migrations will improve policy. Cartesian spatial epistemologies developed on land are ill equipped for the flow and flux of oceans and their species. "Through its complex material properties," the maritime geographers and environmental scientists working with Noëlle Boucquey write, "ocean space resists inscription" (Boucquey et al. 2016, 8). New concepts in management are evolving that are equipped for the ocean's elementality. "Dynamic ocean management" is as "fluid in space and time as the resources and users" it aims to manage (Maxwell et al. 2015, 42). Dynamic ocean management is one approach that develops from the capacities of intimate sensors and is attuned to ocean elementality and maritime migrations.

Humans too are territorial, assuming a soft dominion over shores through pollution, fishing with spears and lines, patrolling in boats and helicopters, swimming, surfing, snorkeling, and scuba diving. Unlike sharks, who can perceive electricity, chemicals, and magnetism, humans fabricate and corporeally graft technologies for these functions, adding lenses and housing to our technologies to perceive and claim more space. Culling systems draw a permeable barrier that differentiates humans from sharks, cutting death out of life while promising the opposite. They are technics of hydropolitical containment developed for terrestrial mammals to dominate the shoreline.

Speaking aquatically, philosopher Gilles Deleuze contrasts the enclosed territoriality of something like a net with a more modular control, writing, "Enclosures are molds, distinct castings, but controls are a modulation, like a self-deforming cast that will continuously change from one moment to the other, or like a sieve whose mesh will transmute from point to point" (1992, 4). Nets concretize the aspirations and failures of a negative and anthropocentric governmentality. In this situation, oceanic elements are exploited, a crude notion of animal agency is manipulated, and convenient human laws are devised. But instead of this being an affirmative biopolitics for multiple species, it focuses only on minimizing the psychological harm felt by humans. It nudges sharks not away from swimmers, but toward their demise. The drone offers the possibility of a modular control—a sieve that allows seawater, sharks, and people to flow into coexistence. Drones present an opportunity to fashion space differently, not through the imposition of the linear cartographies of nets and drumlines but through the deterritorializing drone technicity of atmospheric mobility and automated computer vision. It will introduce new problems but also a suite of possibilities for the preservation of vitality.

COEXISTENCE

The nets distinguish human from shark territory, with severe consequences. As Gibbs writes, "Inherent in these conflicting ideas and action is an assumption that humans and sharks cannot use the same area of water; cannot coexist" (2018, 7). Drones present another way; they offer a blue governmentality that works with the elements, networked technologies, and human laws, but does not attempt to subvert animal agency. In the drone scenario, sharks predate in an elemental territory shared with humans, without tantalizing hooks and unrecognizable nets. Sharks are identified and swimmers alerted. This drone-supported blue governmentality on eastern Australian beaches honors shark agency and its behavior, habitat requirements, and instincts. Drones, in this context, assist us in rationalizing our fears and provide some safety while conditioning us for shark coexistence. The subjectivities of terrestrial beings, too, are transformed by blue governmentality. For Foucault, governmentality shapes choices without fabricating subjectivities and enforcing compliance. Humans are nudged in the direction of living with sharks.

Humans live with biodiversity, in and on our bodies and homes, and in the urban, rural, and wild lands and seas around us (Turnhout et al. 2013, 158–59). Whether we know it or not, many of us today regularly swim with sharks. Videos uploaded by shore-based drone pilots disclose how close swimmers are to sharks, unbeknownst to them, as that proximity can be seen from the elevated vantage point provided by the drone but not from the water surface. Australian drone pilot Jason Iggleden and his Drone Shark App often present videos of sharks near surfers in beaches in eastern Sydney (see www.dronesharkapp.com.au). Iggleden, whose drone has spotted white sharks feeding near swimmers in Bondi Beach and who regularly warns swimmers based on what he sees, advises that "the drone technology is way better than a man up in a tower looking for a fin" (Fanner 2019). Viewers of his app are familiar with the perspective of his drone. Through it, we understand that we swim with sharks. And continue we do.

Frequent beachgoers in Western Australia—open-water swimmers, surfers, snorkelers, and body surfers—know that they cohabitate with sharks and report no problems with the fish (Gibbs and Warren 2014). Many avid ocean users encounter sharks regularly, can identify many shark species, understand the slight risks, and advocate for others to follow suit to learn to live with these predators (Gibbs and Warren 2014). For us open-water swimmers, the sea is an agent (without intentionality) that presents risks and opportunities for sensual engagement with the elements and other species (Anderson 2012). The potential for an intimate experience with a shark is but one of the risks in entering this world.

Practices of living with predators can be inspired by studies of brown bears (*Ursus arctos*), Eurasian lynx (*Lynx lynx*), gray wolves (*Canis lupus*), and wolverines (*Gulo gulo*), all species that are thriving in European countries. After studying the location and population frequencies of these predators, biologists concluded "that large carnivores and people can share the same landscape" (Chapron et al. 2014). These scientists name it the "coexistence model" and contrast it with the "separation model," which divides people from predators (Chapron et al. 2014, 1517). In Australia, we could be informed by these stories of coexistence with predators in densely populated Europe. However, a closer examination of their argument shows a terrestrial bias that is geared toward "novel ecosystems" (Kirksey 2015). The biologists argue that it is possible for predators and humans to live together on farms, ranches, and other anthropogenic landscapes. Here agriculture and pastoralism enhance flourishing for these European predators (Turnhout et al. 2013, 158–59). This is the terrestrial and anthropogenic thinking that ocean/culture attempts to correct.

The ocean offers limitations and possibilities for predator coexistence. Its incongruity with long-term human occupation and its fluid, constant motion mean that submarine spaces are not primarily dominated by humans. Shark nets, baited hooks, and humans swimming by sharks do not constitute natureculture hybridity. So although a "separation model" dividing oceans and cultures is unlikely because of the boundary-corrosive qualities of the ocean—as well as the pervasiveness of anthropogenic molecules, heat, and hunting—ocean/cultures remain a possibility and a goal for marine conservationists.

Drones provide opportunities to monitor the boundaries of ocean/cultures, rather than concretizing them, because they offer a porous and mutable sieve, a "self-deforming cast," in Deleuze's language (1992, 4), for the execution of caring control toward coexistence. Humans will be instructed to leave the sea when alerted to the presence of impressive sharks. Returning to the "coexistence model" (Chapron et al. 2014), this is not a farm that accidentally supports the grazing of roe deer (*Capreolus capreolus*), which are prey for wolves, bears, or wolverines. Theories of coexistence that do not conceive of the elements offer limited, terracentric models that do not transport well into the ocean. Yes, sharks and humans can coexist in the ocean, but because the ocean is not the element where humans exist, a separation between oceans and cultures remains a possibility and objective.

We can coexist, particularly in Australia, a country with a reputation for thriving among deadly snakes, spiders, birds, octopuses, jellyfish, and sharks. The world's first protections for a shark species occurred when the New South Wales state government declared the gray nurse shark (*Carcharias taurus*) a protected species in 1984 (Pepin-Neff 2012). Coexistence requires a willingness by humans to live with sharks, admit the ocean is beyond our control, and have our behavior nudged by drones, AI, and alarm systems. Normatively speaking, the Australian is a citizen relatively compliant and trusting of its government and its technics (Goldfinch, Taplin, and Gauld 2021). Surveys show that Australian beachgoers both accept drones and are concerned about the impacts of the nets (Gibbs and Warren 2014). In reviewing its shark net program, members of the Queensland task force that was established to adjudicate their efficacy admitted that you "can't manage the animals in the ocean but you can influence human behavior through science-based education" (Queensland Department of Agriculture and Fisheries 2020, 3–4). Activist Jonathan Clark declared more emphatically, "shark control program?! I defy anyone to control a shark, one of the greatest animals in the ocean. How do you do that?" (Clark personal communication 2021). Zoologists call this accommodation a result of "mutual avoidance; and mutual flourishing" (Pooley et al. 2017). Coexistence, "even amidst relations widely represented as unquestionably dangerous," such as those with sharks (Gibbs 2021, 11), would be a profound acceptance of interdependence. It would be a humbling admission and collaboration with an ancient foe.

The drone shark programs provide a flight way toward coexistence wherein sharks and swimmers share the shore, with both human and shark being "watched over by machines of loving grace," in the words of poet Richard Brautigan (1968). Unlike the deadly nets, drones enable a lively form of living with—while remaining separate from—the teeth of sharks. With drones persistently monitoring the beaches, nets and drum-lines could be pulled from the oceans. Instead of firm or faux borders sep-arating sharks and humans, the drones would open shoreline boundaries to multispecies and transtechnological movement. The shoreline would be deterritorialized for sharks, bathing humans, and roaming drones.

The drone program represents an acceptance that living with sharks means abandoning the false sense of security that territoriality appears to offer—the illusion that they are over *there* while we are *here*. Con-trolling the lives of certain sharks through catching their bodies in nets is not feasible in a marine world of multispecies flourishing. Instead of faulty territoriality, humans can be trained by drones and alarms to learn to live with sharks and leave when they approach. Swimmers and surf-ers may have to move (and quickly) when alerted by those monitoring the drone's live video feed or by the automatic alarm. It may not always work—drones crash, vision fails, humans and computers err, and sharks have an animal agency we do not understand and cannot control. But this disruption in play, leisure, and—sometimes, rarely—life itself is a necessary sacrifice for the "ecosystem services," if you must, or opportu-nities for multispecies flourishing, if you will, that are provided by this apex predator.

8

ENDING

CORAL/CULTURES

UNDERWATER DRONES IN A MARINE PARK

The human niche is niche switching. Humans explored all continents and settled in most. But some places remain inhospitable—the bottom of the sea, the upper reaches of the atmosphere, and the poles. Extreme ecologies—the thickest forests, the driest deserts—are uncomfortable for many without acclimatization. Some disagreeable lands are so dominated by other organisms that coexistence is challenging. In January 2022, I, my wife Robin, and our daughter Io arrived in one such locality where birds, insects, and reptiles rule. Lady Musgrave Island in the Capricornia Cays National Park in the Great Barrier Reef World Heritage Area, Australia, is billed by tour agencies as a lagoon of turquoise waters, white sand, and colorful reefs, just two hours' cruise from the mainland— perfect for an adventurous day journey. This is certainly true if you stay around the shore. But off the beaten path, in the dense pisonia forest (*Pisonia grandis*) of the island's interior, is a rookery of aggressive black noddies (*Anous minutus*), bridled terns (*Onychoprion anaethetus*), and mut-tonbirds (*Ardenna tenuirostris*). These birds have no fear of humans. Cling-ing to every tree crook are nests formed from their calcium-rich guano. The trees drip with bird droppings and shiver with the clatter of chicks. Sharp-billed parents swoop at passersby through the speckled light of the broad-leafed canopy. The floor is littered with dead chicks—sticky seed clusters suffocate adolescent tail feathers, pulling the chicks into the feces. Bird wings and sundry former organs squish underfoot and, against the white sand, form a fertile matrix for future offspring of the appar-ently carnivorous pisonia trees. This has been going on for a while. Unlike her relative islands, Lady Musgrave doesn't migrate—the bird shit, dead bodies, and tree roots lock the shift of the island's sand to the seafloor.

From the lagoon and into this atmospheric cacophony and earthly muck we ventured with seven wheelbarrows of gear to our campsite—booked fifteen months in advance. We arrived a few days later than planned—Cyclone Seth made the swell and winds too high to sail. The captain warned us we would be "climbing mountains" to get to the island on this day after a tropical storm; a majority of those we portaged with revisited their breakfast in several "convenience bags" as we powered between a sandbar and a standing wave and ascended littoral hills. Stoi-cally, we three made it without stomach spillage. With the travel behind us, we had a real issue to solve. We needed a shelter from the falling baby birds, poop, and insect larvae. Immediately. We quickly strung up a tarp from four trees; the droppings, seeds, and screeching chicks regularly drummed this elevated guard as we made our tents, hammocks, and a kitchen underneath. We will acclimate to the sounds, smells, and sights of this avian-ruled world, we promised ourselves. We came here to con-duct a drone survey of coral bleaching, we remind each other. Over the course of the next two weeks, I will fly slowly across sections of the coral reef and sea cucumber patches at a height of thirty meters, collecting images that will be restitched together to create a highly accurate ortho-mosaic map of animal activity. This map will be compared longitudinally with previous and future records to help those interested better under-stand the direction of entropy on this island.

But upon arriving, it soon became apparent that seabirds and coral are just two of the charismatic clades that leave a visual trace for the drone to index from above. On the first night, we met another. At sun-set, noddy and tern parents return from fishing and the chick chatter reaches a fevered pitch. After feeding, the caterwauling subsides and another bird comes to rule the soundscape, the short-tailed shearwater, or muttonbird—a name given to it by shipwrecked sailors who, having first extracted its flavor-befouling gland, enjoyed the bird's lamb-like taste. Muttonbirds stalked our nights. They would clumsily fly to their nearby underground nests, crashing into our tents, tarps, and bodies— once knocking me in the head, the toothbrush flying out of my mouth and into a pile of feces. Like a chorus of mourning children, the muttonbirds would howl from their subterranean hovels. Good sleep was impossible.

Adding to this immersive experience were green sea turtle (*Chelonia mydas*) mothers emerging from the sea to lay their eggs. One night, a 150-kilogram turtle greeted my tent and me inside it. Apparently, I was sleeping in the ideal place to dig and lay eggs. It proceeded to claw over the tent with its long paddles. Failing this or responding to my cries for mercy, it began to burrow under my tent. There was no reasoning with this maternal reptile exploring for a nest. I respectfully declined its advances by relocating elsewhere. I dragged my tent behind a block-ade of chunks of coral and fallen timber. The next night, I restrung my hammock on a small coterie of she-oaks (*Casuarina*) nearer to the shore. That evening, a hawk moth (*Sphingidae*) outbreak or flush began, as they do after tropical storms: the insects emerge from the guano and bird-body fertilizer. The moths blackened out our little solar lights. Blacktip reef sharks (*Carcharhinus melanopterus*) patrolled the evening shallows, searching for turtle hatchlings. Muttonbirds moaned. And in the morn-ing, the cries of the terns began anew. This is a place, like so few today but so many before, that humans have not domesticated with pesticide, deforestation, overfishing, and resource extraction. Many years ago it was tried here on Lady Musgrave Island: species were introduced, build-ings erected, and an industrial turtle fishing operation briefly functioned to produce an inedible stew. It was abandoned for lack of a market. Today different marine and tourism management strategies govern the island and human access.

Lady Musgrave is closed to visitors for several months a year; recre-ational fishing is limited; and spearfishing is outlawed. As a protected island, it is part of the 2.8 percent of the ocean with this status. State con-servationists closely monitor the health of its coral, watching and elim-inating coral predators like crown-of-thorns starfish (COTS, *Acanthaster planci*). Settler colonists like myself can take fish by line (but not by spear point). One of the seventy or so Aboriginal and Torres Strait Islander people who gathered and hunted here called this island Wallaginji, or "beautiful reef," and they can fish, collect, hunt, and care for their traditional sites in and around the Great Barrier Reef (Dobbs 2007). I met several Indigenous Gidarjil Bundaberg Land and Sea Rangers who were here to monitor the island and their Sea Country. They care for the seagrass, mangroves, and saltmarsh on the mainland and the coral reefs offshore. They use dogs to find introduced red foxes (*Vulpes vulpes*) that kill turtle clutches. They told me about their ancestral connection to the sea and its inhabitants. They were excited because the previous month, the Gidarjil Development Corporation had received a substantial grant to develop aerial and underwater drones to better connect to Sea Country. Here customary knowledge, settler-colonial science, and elemental technologies collaborate to manage an island. This is a protected area that embodies ocean/cultures; humans are here temporarily to care for the land, leaving to let it be.

Other caretakers also populated this little atoll. The next evening, I was awoken from my she-oak swing by the dim light of old-fashioned flashlights. Herpetologists were using a lumen of light below turtle perception and conducting a survey of nests. I joined them, and for several nights we counted, sized, and tagged sea turtles and the rare loggerhead sea turtle (*Caretta caretta*) late into the night. One moonless evening, as the lighthouse scanned the horizon in a racing glow, the resident turtle scientist, Jim Buck—who with his family has been leading this turtle survey for nearly thirty years—told us a story of care and control.

One green turtle mother had mistimed her return to the sea with the receding tide. She was marooned on the flats of the coral rubble, unable to get to the ocean, and the sun and heat were rising. Jim nudged her along, using her fear of him to motivate her toward a small pool of saltwater where she could safely wait for the next ascending tide. Her paddle gently bled a red cloud into the saltwater. Witnessing this brutal encouragement, a tourist accosted Jim, demanding he stop and leave the turtle alone. Jim softly told the man he would relent after he helped him pick up the reptile and place it in the nearest pool. Otherwise, it would overheat and die. The man agreed and, despite her immense weight, they were able to scurry across sharp coral and deposit the animal in a puddle where she could cool herself until the tide overcame her. Jim's science and conservation overlap, with ocean and culture meeting in acts of long-term care and temporary control.

The next morning, a uniformed official traipsed through our campsite spraying a red-stained poison on exotic weeds like crowsfoot grass (*Eleusine indica*). I asked Jim about this, and he reported that this island, like many others in the Great Barrier Reef, is undergoing a rewilding—a process of returning it to an earlier status. Rewilding requires tactics, technologies, scientific epistemologies, and local knowledge. In this reef, COTS, which eat coral, are killed by divers with a poisonous injection. On other islands, Jim and his colleagues use an industrial steamer to kill weeds. Years ago, goats, left as potential food for anyone cursed to be shipwrecked here, were herded and slaughtered before they ate all the island's precious foliage. A small resort was dismantled, its concrete foundations now an obstacle for nesting turtles. On nearby Lady Elliot Island, they felled foreign trees and moved a crested tern (*Thalasseus bergii*) rookery away from the airport runway. For the traditional owners as for Jim, conservation is hands-on—monitoring coral and seagrass, carrying turtles by their shells, poisoning weeds, felling trees, killing goats and COTS, puncturing thick reptile skin with numbered titanium tags.

Resilience is the goal, Jim tells me one day as we wait for the composting toilet, the only building remaining on the island. Older relations—like those between Aboriginal and Torres Strait Islanders and Sea Country, and between the pisonia trees and the nobby chicks entangled in their seeds and sap, and the turtles and the sharks—are antifragile and therefore more resilient than the "emergent ecologies" of goats, crowsfoot grass, and capitalist tourism. Humans, with conservation methodologies and technologies, can be caretakers of complexity, capable of monitoring and intervening, but also able to leave the island to its ancient ways.

The nexus of care/control was graphically felt by our nine-year-old daughter Io that evening as we witnessed our first sea turtle hatching. Night after night, sometimes with more intimacy than we humans desired, we watched mothers crawl from the sea and up the slope of coral rubble toward the loose sand below the pisonia boughs. There they would dig for hours—first with their front paddles, followed by their back—a hole bigger than themselves and into which they would deposit their eggs. This night, some of those eggs hatched and out clawed tiny, adorable sea turtles that struggled to descend to the sea. Io was beyond herself with concern for their struggle over the smallest bit of driftwood. To alleviate her and the turtles' stress, she carved a smooth canal out of the sand, a pathway they could take to get to the ocean (figure 8.1). Some diligently followed her easing route. Others pioneered a path over and out of her furrow. All of the delicate reptiles made it to the sea, where white-striped

FIGURE 8.1. Io modulating the direction of green sea turtle hatchlings, 2022. Photo by author.

octopus (*Callistoctopus ornatus*), black-tip reef sharks, Eastern reef egrets (*Egretta sacra*), and silver gulls (*Chroicocephalus novaehollandiae*) captured and feasted on them. Io and the other children on the island ganged up on these cephalopods, fish, and birds, chucking chunks of coral in their direction. Some dropped their turtle catch; most did not. Parents were overheard consoling children with "it's nature's way."

The next morning, I think about this as I prepare for a drone flight to document the health of a northern section of the reef. From the return of Indigenous caretakers to this island and the work of Jim Buck and his volunteers, to the rewilding efforts of the conservationists working for the Queensland government, and the tiny trough made by my daughter and her friends for the turtle hatchlings—such efforts of care feature acts of control. Technologies from titanium tags to the shells Io used to carve a path to the sea extend human care, the efficacy of which is balanced by elementality and animality.

With this in mind, I lift off the soft sandy soil littered with the tracks made by green sea turtles finding and digging nests. During these months, between fifty and two hundred turtles a night come out of the sea to this island (figure 8.2). Their pathways leave an evolutionary trace of how they abandoned terrestriality for marine mobility but retained a need to lay eggs on land. When not on land, turtles—like whales, seals, and most corals—live much of their lives in the first surface meters of the ocean, and therefore are seeable by drone. In the days that followed,

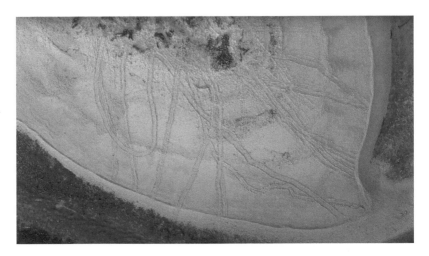

FIGURE 8.2. Green sea turtle tracks by drone, Lady Musgrave Island, 2022. Photo by author.

I made orthomosaic maps of turtle nesting pathways, sea cucumber densities, and coral reef colonies, and I followed black-tip reef sharks and green sea turtles around the lagoon. These encounters with wildlife were mediated by drones, but others are made possible by different techniques of blue governmentality—the orchestration by the Queensland government for the return of Traditional Owners, the funding of the work of scientists like Jim Buck, and COTS eradication efforts. Other blue governmentalities include the regulations of the Great Barrier Reef Coastal Marine Park, which allowed me and my family to camp on the island and follow strict orders on what can be done here—no fires, and only five shells can be collected (I later discovered in Io's luggage that she violated this ordinance by a factor of ten). All water, food, shelters, and trash that were brought in had to be consumed or taken out. These rules are designed to keep the island in a state of ocean/culture —with the right amount of human engagement to support yet retain resiliency.

Ultimately, we were the ones who proved to be fragile. Cyclone Seth had subsided. But during our stay it had migrated out to sea, where it found more heat and force and returned to Lady Musgrave. Folks on the ship that kept us supplied with fresh water, ice, food, and cold beer— something to take the edge off of being the unwelcome guests in a rookery—and other necessities told us that they would suspend their service for five days as the storm developed and passed. The swell, wind, and waves would be too substantial for their ship to overcome. We were welcome to wait it out indefinitely with the limited supplies we had. We decided the safest move would be to pack up and depart, leaving the island to the nobbies, turtles, and sharks. Wind and water conspired to put an end to the fieldwork for this book.

While conservation technologies provide one aperture for seeing the care/control nexus of ocean/cultures, the ethos of marine parks such as the Capricornia Cays National Park that covers Lady Musgrave Island offers an exceptional example of the human engagement to liberate oceans from cultures. According to marine biologists, to defend biodiversity, at least 26 percent of the ocean should be protected (Jones et al. 2020, 188). This will require much more than drones and will demand the full suite of blue governmentalities, including no-fishing or no-take zones and the regulation of the high seas. The United Nations Convention on the Law of the Sea, the first-ever high seas conservation treaty, which was being debated at the United Nations at the time of this writing, may take a step toward this monumental goal of protecting over a quarter of the ocean (Fox 2019). It is yet to be seen if this law will be enacted and how well it will be enforced.

But academics who are skeptical of parks, wilderness, and nature can take solace in the fact that only 2.8 percent of the ocean is currently fully or highly protected (IUCN 2013; Marine Conservation Institute n.d.). It is true that the rights of folk to take fish legally and customarily might be hampered in the implementation of marine protection areas. Fortunately for the Gidarjil, they can acquire a Traditional Use of Marine Resources Agreement (TUMRA) that will certify their taking a limited number of sea turtles for cultural purposes in waters surrounding Lady Musgrave (Great Barrier Reef Marine Park Authority 2018). So around 97.2 percent of the oceans will continue to be free from what some academics claim to be the discriminatory tyranny of parks that protect biodiversity. Messy oceancultures still rule over the imperfect blue governmentality of ocean/cultures.

Of course, this absence of protection is not actually the good news for coastal and fishing communities that these academics claim it to be. Protecting the oceans from illegal fishing, overfishing, dynamite fishing, bottom trawling, offshore oil drilling, and industrial development rehabilitate their marine food- and resource-providing stocks. Population densities, biomass, diversity, and size of fish are all higher inside marine parks than outside (Halpern and Warner 2002; Gill et al. 2017).

Parks mean more and larger fish, which may leave the park and be caught by regional subsistence fisherfolk. It is an anthropocentric framing of their possibilities, but parks are better for hungry families who live on park borders.

Nevertheless, it is important to consider the implications of this positive biogovernmentality. Some Thai living on the Andaman Coast, for instance, claimed a recent marine protection area had a negative impact on their lives. Others from the community disagreed, noting that because small-scale fishing was allowed and enforcement was lacking, the marine park designation did not hurt their livelihoods (Bennett and Dearden 2014). After interviewing eighty-five local fisherfolk, the marine social scientists concluded, "MPAs [marine park areas] have the potential to conserve the environment and increase fisheries while contributing positively to social and economic development in local communities *if* (a) local development considerations are taken into account and (b) they are effectively managed and governed" (Bennett and Dearden 2014, 114). In a hierarchy of needs, food is more important than freedom, and temporary cultural inconveniences for a few humans might be necessary to protect the prosperity of many more future humans as well as entire marine ecosystems. The control executed by marine parks means caring for marine beings, human and otherwise.

The UN's goal is to protect 10 percent of the seas. But the globalist threat to individual freedom perceived by several academics and posed by "fortress ocean" continues to limp along with many loopholes for fishing and lackluster enforcement (Brockington 2002). Thinking elementally, a marine park is much less a fortress than a terrestrial park, which might prohibit motorized vehicles, exogamous cattle, or subsistence hunting. In both situations, the boundaries of the so-called fortress are permeable by atmospheres, people, and other species, but this is ever more so the case with the ocean, given that its molecularity is fluid, its chemistry hypercorrosive, and its biology largely migratory. Within these elemental dynamics, it is difficult to measure the efficacy of parks for biological diversity and the effects that the park might have on the livelihoods of locals. Other factors are depleting and changing the ocean's biomass—heating, fishing, acidification, pollution, and so on—all of which flow into marine parks affecting local maritime cultures. Despite the best intentions and surveillance technicities in and around marine parks, blue governmentality—like its green kin—is a biocentric aspiration, neither a clear possibility nor a discriminatory achievement.

Reefs are home to a quarter of the world's fish and sea creatures, and they provide 17 percent of the animal proteins eaten by humans (FAO 2014). As is well documented, corals are struggling worldwide because of overheating, acidification, bleaching, dynamite fishing, dredging, overfishing, chemical pollution, and crown-of-thorns starfish (COTS) (Heron 2018). Lady Musgrave is one of the islands undergoing a program to minimize COTS. Throughout the year, divers spill out of crafts plying the waters around Lady Musgrave, armed with syringes full of poisonous vinegar to inject into and kill the coral-eating starfish. From my limited week-long exploration with both aerial and underwater drones and swimming to survey a small section of the reef, they may be effective. I did not see a single COTS. I would have been provoked to pull out my dive knife and give it a fatal blow—while avoiding its many toxic spines—if I had the chance to control the COTS and care for the reef.

Professor Matt Dunabin has invented RangerBots, underwater drones that use AI to identify COTS and then inject them with a toxin, killing this rampant predator of the Great Barrier Reef (Gartry 2018). Current COTS removal strategies are manual. Two teams of ten to twelve divers, rotating ten days on and four days off, dive, cull, and remove the starfish. Since 2011, roughly 300,000 have been eliminated (Australian Government n.d.). Legal and ocean scholar Irus Braverman questions the morality of this hunt, balancing the killing of one sea creature to save another: "At a cost of three cents per COTS injection, this is possibly the cheapest animal death ('removal'), and, by extension, the cheapest animal life, on earth" (2020, 149). The paradox is that these drone and manual interdictions kill COTS to save coral. For Braverman, this is a compelling account of the conundrum of robotic management of the ocean. Drones that do the dirty, dull, or dangerous work of killing, according to her critique, eliminate the need for humans to look deeply at what or why we are killing at all. Supporting this critique, is, unsurprisingly, linking RangerBots to military drones—both tools that afford the ability to kill, but from a distance.

One afternoon we left Lady Musgrave Island to venture out to the reef edge in our three-person inflatable kayak to descend a Chasing Dory underwater drone to the seafloor in a channel where we had previously swum with a gang of massive manta rays (*Manta*). We quickly discovered that floating on an inflatable kayak in rolling waves, piloting an underwater drone that you cannot see—and that offers a limited aperture for computer vision and limited feedback on where it is—is quite a challenge. Finding and navigating toward a COTS to kill it—even with AI to assist—would be insurmountable. This humble experiment with a small underwater drone proved that such techniques interface less with submerged fauna than they do with the elements. Similar elemental challenges are likely for RangerBots, should they ever be seriously deployed.

Braverman's approach is typical of drone critique, which starts with drones as military weapons and does not account for the engineering history and culture, application, materiality, and functionality—the relative technicity—of the conservation drone. Underwater drones erase and displace humans from the governance of life, she argues (Braverman 2020, 153). I found the opposite. Drones invite humans to confront death and face extinction. Drones bring scientists, activists, and threatened species into intimate states of entanglement, where humans confront the task of triaging one life for another. They approach these uncomfortable realities through their tools, which amplify their care and control into existential and elemental spaces. This is difficult, technical, and compassionate work on behalf of broader ecological goals in brutal underwater worlds.

RangerBots and their kin will need to solve problems related to electrical power, computer vision, and communication if they are going to improve—and draw nearer to the dystopia predicted by conservation critics. The ocean's elementality is a major problem for clear seeing, a potential source of power, and a mediator for communication. For example, to improve AI identification under the sea, MIT oceanographer Derya Akkaynak developed Sea-thru, an algorithm that removes the water from underwater photographs, allowing an element-free depiction of coral reefs (Djudjic 2019). The elements are not only disruptive to accurately sensing the sea but can empower it. Getting around the problem of powering underwater sensors is the work of other MIT scientists who have devised a way to generate electricity by absorbing and reflecting sound waves through the water (Matheson 2019). Penetrating the sea with information and directions from the surface continues to limit remote control and data retrieval from underwater sensors. Fadel Adib, an assistant professor at MIT, is developing a system that translates water-based sonar into atmospheric binary code at the ocean surface. He said, "Trying to cross the air-water boundary with wireless signals has been an obstacle. Our idea is to transform the obstacle itself into a medium through which to communicate" (Matheson 2019). Like the present, the future of conservation addresses elementality with relative technicity, finding ways to creatively work with, through, and around the elements.

Like the RangerBot, the underwater drone I directed through coral corridors is not a military drone, nor is it a highly effective tool for control. Its batteries need to be recharged, its software updated, and its hardware maintained, and thus its reach is far too small to cover much of the 2,500 kilometers of the Great Barrier Reef. The RangerBot is under the same constraints and offers similar limitations. It is an experiment to care for the thriving of a small web of creatures that constitute a fragment of a coral reef. Instead of prudently admitting this limitation, Braverman instead casts underwater conservation drones as ominous portents for "planetary governance" (2020, 148). If this is the future of "robotic management not only in the ocean, but also in planetary governance writ large" (148), then our oceans are truly ungovernable, and our robotic overlords are far in the future. Views of negative biopolitics such as this magnify discrete and often prototypical examples into universal warnings about technology, surveillance, and governmentality.

This technophobic criticality plays on hypotheticals about the loss of Western liberal values such as privacy and freedom. Blue governmentality is an aspiration, not a reality. Its partial consequences can be many—surveillance and grammatization of the ocean, for both humans and nonhumans—but implementing blue governmentality is an ethical quandary far more profound than whether COTS or coral are more valuable. It is essential to remember that conservation is not for humans but for other living beings. What may bother Western senses of privacy and freedom does not have the same effect for the organisms that make up coral. Not to be reductive of their rich underwater *Umwelt*, but physical well-being is more important for these zooxanthellae and invertebrates. Answering these puzzling questions of what to kill (and how) in the affirmative is necessary for the emergence of ocean/cultures, where cultural processes facilitate oceanic resiliency. At the same time, the efficacy of this relative technicity will be confounded by elementality.

This book has identified and interrogated technicity and elementality, the baseline factors that are indispensable for understanding the potentials of drones for conservation purposes. It has examined how they impact the formation of multispecies intimacy at the moment of embodied proximity. It offers blue governmentality to understand how relative technicity and elementality confound human political power. The goal of drone oceanography is to foster ocean/cultures: oceans whose wildness is supported by technologies.

Originary technicity is the evolution of technology as an externalization of human bodies. It features the grasping capacities of the hands, the vision of the eyes, and the mobility of the feet in gathering and hunting operations—the "whole of our evolution has been oriented toward placing outside ourselves what in the rest of the animal world is achieved inside by species adaptation," wrote archaeologist Andre Leroi-Gourhan ([1964] 1993, 235). Originary technicity is a theory of technological emergence from the human drive to externalize, extend, and amplify sensing and movement proxies outside our bodies and into other worlds. Technologies, cognition, bodies, and cultures evolved collectively to overcome physical and geographical limitations on movement, sight, and the execution of force. The fitness of originary technicity modified our bodies, changed our brains, and enabled cultural complexity. As technologies evolved, so did we.

The sensorial extensions needed for data gathering in scientific and activist operations manifest this ancient unfolding. Ships carry bodies across seas to geographical and biological anomalies. Telescopes and microscopes tunnel eyes nearer distant and tiny others. Submarine arms collect underwater specimens. Helicopters elevate and carry body and mind over swell sets. Likewise, the drone's relative technicity of speed, movement, grasping, programmability, and force from afar exploits the atmosphere akin to how ships and submarines utilize the sea and the helicopter the air—as a platform for discovery, intervention, and control. The expansion of drones to and below the sea surface illustrates how the elements are the substances through which technicity is expressed.

Ocean drones lengthen these drives into the element, intra-acting with whales, porpoises, sharks, seals, seabirds, turtles, sharks, and COTS as well as with fisherfolk who might harm them and swimmers who might be harmed by them. Conservation drones liberate specific humans' interior milieus—in this context, the mix of scientific, ecological, and practical sensibilities that constitute nature realism—into the elements, which become entangled with actants in exterior milieu: organisms, technologies, and laws. Related to touch, relative technicity enables novel forms of intimacy and leads to the collection of robust data about unsettled marine life.

Chapter 2 began by gazing at whales from shores and ships, mythologizing their origins, capacities, and purposes (Kinzelbach 1986; Papadopoulos and Ruscillo 2002). Cetologists studied the gruesome entrails of harpooned whales, shot them with crossbows and tags, and floated above them with hydrogen balloons before flying petri dishes through their exhale on drones. With each step, scientists got higher-resolution data about whales, procuring vibrant matter from their whalehood (Angier 2010), until the arrival of artificial intelligence automated the identification of their flukes, tracking their movement through space and time (Gray et al. 2019). From imagination to imaging, the drone's relative technicity has brought humans and whales into closer, living contact. From hearing recordings of humpback songs, to measuring the importance of their feces for the oxygenation of the atmosphere (Roman and McCarthy 2010), to watching them care for their young via drones—sensing technologies have shifted cetology from necropolitics to zoēpolitics. It has always been intimate—either wading through the grax of a gutted whale in search of a tag or peering down at them from a drone. With the drone, intimacy is with the living instead of the dead whale.

How technology increases interspecies intimacy is also evident in chapter 3, which traveled with activists as they developed and deployed a range of elemental technologies—from seafaring vessels and prop foulers on the ocean surface to net-ripping installations on the ocean's bottom. Taking to the air to heighten their reach, the activists fired water cannons, pie filling, and methylcellulose to contest whale harvesting. First helicopters and then drones became essential tools for dilating visual reconnaissance over the sea and venturing closer to marine animals. This capacity materialized in capturing evidence of illegal shark fishing in chapter 4, and hopefully saving sharks from fatal nets in chapter 7, while chapter 6 made evident the mortal consequences of relative technicity's direction or, in this case, misdirection of force from afar, resulting in an entropic fate for seabirds and orcas. Here, falling drones and abandoned elegant tern nests shared a fatal embrace.

Originary technicity is a nonlinear evolution: its motion into the adjacent possible remains open to bifurcation toward necropolitics and zoēpolitics, death and life. Control is imperative in either. And while control over living and dead bodies has been the overwhelming trajectory of originary technicity, the capture of information is an increasingly important direction for human evolution. Information and resource acquisition have always been interlinked. "Operational memory" and "programming," once stored in practices of orality and performativity, are today relegated to computers, wrote Leroi-Gourhan ([1964] 1993, 238) long ago. The drone extends mobility, vision, and force from afar as well as programmability to the ocean. This programmability, while balanced with the care of applied science and marine activism, is partial in its perspective as well as its application. Drone conservation complements other aspects of blue governmentality—the analysis of whale snot, the storying of seal struggles, the prosecution of shark finners, the legislation of drone shark surveillance, and the establishment of marine parks. Ideals of conservation meet the loopy determinism of a technicity that has been made relative by the affordances of drones and their elemental contingencies. The ocean and its inhabitants remain unprogrammable, at this point, with this technology. A tension between technological idealism and determinism plays out over these troubled waters.

ELEMENTALITY

The drone is a technology that emerges from the interior milieus of scientists and activists. It amplifies senses and bodies and cocreates imbroglios of humans and nonhumans. Lofted by the atmosphere and lifted above the sea, the conservation drone mediates science and activism, monitors endangered species and poaching, and attempts to govern marine flourishing. The elements—the saline and aqueous ocean, the buoyant and whirling atmosphere, and the storms they produce upon contact—affect the drone's flight and therefore impact drone conservation. Lofted and floated drones generate proximity between conservationists and species, enhancing a closeness that summons an ethics of care.

In field trials of technological prototypes, scientists often document elemental conditions. Time, temperature, wind, and glare are all dutifully noted. But while the ecomaterial turn in media studies and environmental humanities has encouraged scholars to attend to nonhumans, it has often meant an emphasis on living nonhumans. Addressing this deficiency, the agencies of nonliving elements in the realms of meteorology and geology have been positively identified (Povinelli 2016), and the elements have become a focus in environmental and material media studies

(Parikka 2015). The result of this scholarship is a world that is not only livelier, but also less anthropocentric.

Carrying oxygen, suspending bodies, delivering nutrients, and providing other life-sustaining functions, this planet supports life because of the qualities of its atoms, molecules, and elements—oxygen for firing respiration, water for lubricating cells, the sea for suspending lifeforms, and the atmosphere for communication. Chapter 2 finds the elements mediating communication between pilots and the drone, the electromagnetic field in the atmosphere enabling the drone's technicity to stretch toward the whale's breath. The whale's exhale is elementally lifted as a warm vapor for the drone to collect and return to the raft of gas-inflated pontoons floating on the sea surface. In chapters 3 and 4, we witness the numerous ways the sea and air, despite their chaotic mixtures in storms and waves, make possible the drone's technicity and the work of marine activism to chase whale and shark poachers.

The elements also do the opposite, challenging life. Poisonous gases are emitted from volcanoes and battlefields; suffocating waters sink swimmers and sailors; the weft of the Earth's gravity pulls bodies to the ground, grinding our bones as we age. Chapter 5 provides insights into what can be (or is not) done with the data collected by the drone's technicity. Virtually indestructible in the face of elemental pressure, the sea-surface drone tracked seals and journeyed to the edge of sea ice, only to provide information for "basic science"—not for the articulation of seal struggle. Chapter 7 details the effort to stop shark death by replacing deadly nets with sea-monitoring drones. The elements are central to this proposition. In the opaque and quiet waters of Queensland's north, shark barriers like stiff fake kelp forests are suitable options. In Queensland's south and New South Wales, where the waters are clearer and more forceful, drones are a contender to replace the nets and baited hooks and prolong shark flourishing. Piloting error and animal agencies are key factors, but the conservation efficacy of drones is contingent upon what the elements afford. In this manner, chapters 5, 6, and 7 make evident not only what the elements invite but also how they limit mobility, stifling sensing under the sea. In many cases, it is a matter of life or death.

In *Oceaning*, the term *elementality* refers to the conditions of the air and the ocean and how they impact sensing. The materialities of the atmosphere and the viscosities of the ocean affect technological praxis. The elements make drone oceanography possible by generating lift and suspension, but also disrupt it through gusts, wave action, glare, and fog. Vitality is also elemental, thriving on the metabolic chemicals coursing through the air and ocean, sedimented in soil, and running across the hard Earth. Elementality is existential mediation—the elements carry and subvert messages, intimacies, and life. Sentience on this planet is one of elemental flowing—as Japanese Zen master Dōgen reminds us from the thirteenth century, "all mountains ride on clouds and walk in the sky" (1995, 141). As philosophers from Baruch Spinoza and Gilles Deleuze to physicists Carlo Rovelli and Brian Greene stress, there is no being, only becoming (Deleuze [1970] 1988; Rovelli 2018; Greene 2020). Life's fluidity is carried, interpenetrated, and mirrored by the elements' movements.

INTIMACY

Originary technicity eclipses the space and time between actants. The elements bring together as well as rip apart. This is as it always has been for our species and for the evolution of technologies. Prehistorically, one important objective of innovative technology was to widen human agency in the deadly direction of the calories contained in other species. Some of the first technologies—throwing sticks, nets, snares, spears, projectile points, and arrows—were designed to collapse space, exploit the atmosphere, and slow the pace of escaping food. Anyone who has shot, trapped, and butchered an animal understands the importance of knives and the stages of intracorporeal intimacy that killing and rendering make possible and that feeding requires. Life demands interspecies intimacy, with both the living and the dead.

Drones are commonly understood as remote-sensing technologies, but a more accurate description is that the drone is an intimate sensing system (Helmreich 2009b). Oceanographers may not recognize it as such (marine activists, motivated by passion and empathy, might be more willing to do so), but drone technicity and elementality generate the conditions for multispecies intimacy. By overcoming distance between subjects, allowing higher-resolution witnessing, and inviting a corporeal grasping, the drone's technicity generates a felt and physical proximity between humans and nonhumans.

Chapter 2 indulges in an exceptional example of this embodied intimacy, as the inside of a whale is distended through its breath, and the contents are scooped up by a drone. In chapter 5, sealhood is temporarily disturbed: mothers separated from pups, blood drawn, odd haircuts given, surveillance devices mounted, and predation and prey tracked by drone. This intimacy between people scientizing with whales and seals is like that proposed between swimming people and sharks. In a speculative chapter 7, the nets that kill and fail to separate humans from sharks are abandoned, and drone surveillance enables the living bodies of sharks and people to intermingle outside the shoreline break. Likewise, chapter 4, with its examination of shark fin operations, and chapter 6, with its look into the isomorphics shared between falling drones and collapsing species, expose the fatal intimacies that technologies and elements afford. The activists in chapters 3, 4, and 7 each exploited that fierce intimacy, returning with images of finless, lifeless, and floating sharks and gutted whales. The goal of these images, sometimes graphically contrasted with the fragility of the activist body amid danger and death, is to widen the affect of this intimacy through social media to politicized communities.

In chapters 3 and 4, drones were flown not toward living marine species but at those who would do them and other species harm. A disparate form of intimacy was generated in these direct actions, one that situated activists and poachers in proximity. The affirmative intimacies that drone conservation hopes to achieve are oriented not toward this demise, but toward mutual, multispecies flourishing. In this process, multiple elements are crossed, several communication technologies converge, different vessels are engaged, novel forms of atmospheric activism develop, and an embodied politics of sensing emerges (Vertesi 2012). Data intimacies hail the drone to become an existential technology for the protection of endangered life. As chapter 4 explored, reciprocity may be achieved through law and prosecution or, as chapter 5 demonstrates, through the narrating of multispecies lifeworlds (Oppermann 2019; Van Dooren and Rose 2016). As it attempts to facilitate conservation, the drone makes evident the limits of law, prosecution, and punishment—and the persistent persuasions of corruption.

Originary technicity and elementality generate the spatial, embodied, and sensorial conditions for this intimacy. Care and responsibility may follow this closeness. And yet meaningful, measurable care requires a degree of control. It is toward the positive biopower of blue governmentality "that the intimacies of mediation that hold us together come to be seen not as forms of distance but as forms of closeness" (Bratton 2021, 70). This brings us to one of *Oceaning*'s key contributions: developing a theory of care/control within the elements and on the edge of technology: blue governmentality.

BLUE GOVERNMENTALITY

Originary technicity and elementality create possibilities for interspecies intimacy. The care that may follow can begin with basic science, data collection, and prototype testing. It can be explicit in direct action, surveillance, evidence gathering, storying, interdiction, policing, prosecution, and legislation. Care necessitates control. Ultimately, the objective of conservation is to control population frequencies. This biopolitical control might manifest through discursive interventions in the realm of text, voice, and visual media, or in more physical efforts of keeping something alive or restraining life-threatening activity. Throughout this book, aspirations for control are tempered by the animal's own agencies, the nonlinear dynamics of oceans and atmospheres, and the limitations of legal mechanisms. Blue governmentality encapsulates these care/control contingencies of originary technicity, elementality, and human interventions. Implementation is difficult, but *Oceaning* is evidence that the drive exists for a limited yet affirmative governmentality. The design theorist Benjamin Bratton depicts these mechanisms for a pro-life

> *biopolitical stack*, if you like, an integrated, available, modular, programmable, flexible, tweakable, customizable, predictable, equitable, responsive, sustainable infrastructure for sensing, modeling, simulation, and recursive action. Climate science has all of these except the all-important recursive enforcement part. As of yet, it cannot act back upon the climate that it represents, but it must. Just as the medical model does, it must not only diagnose but also cure. For any positive biopolitics to be possible, we must realize that the issues we face are due to an absence of control over what matters, and an excess of investment in things that do not. (2021, 145–46)

When networked with its technological kin, the conservation drone and its expanding set of features and back-end processing is one of a suite of mobile sensors capable of contributing to the formation of this biopolitical stack. In offering an affirmative biopolitics of blue governmentality, *Oceaning* examines the spectrum of the use of force to protect endangered marine fauna. The activists in chapters 3, 4, and 7 enact the affirmative biopolitics provided by drone technicity and elementality to protect whales and sharks. By tracking and catching poachers and documenting the brutalities of shark and cetacean killing, they enforce a multispecies justice that they believe others have not. This affirmative blue governmentality brings together the surveillance capacities of drones with vigilante justice to externalize the nature realism of biocentrism. It is understandable that the conservation drone's adoption from its military cousin of the "all-important recursive enforcement part," in Bratton's wording, does not sit well with many critical scholars. We can take comfort in the crash theory of chapter 6, which revels in the foibles of technology and facts of entropy. It is not going to work as planned. Tradeoffs will be made, unintended consequences will be endured, and failures will proliferate.

And yet, a truncated blue governmentality can emerge. Chapters 2, 5, and 6 reveal the work needed to defend marine flourishing. With the whale blow DNA unexamined, the seal data unstoried, and drones falling from the sky, disrupting terns and potentially orcas, with legal impunity for the pilots and questionably effective robot starfish killers, the capacity of blue governmentality to enact care and enable thriving is tempered. An abundance of care and the fault lines of control characterize these cases. While technology, elements, and animal agencies are factors, the efficacy of blue governmentality is also contingent upon the enactment of laws, regulation, management, policing, and enforcement.

Blue governmentality offers a compromise. For philosopher Roberto Esposito (2013), affirmative biopolitics is defined by the tensions between *communitas*, or what binds beings together, and *immunitas*, what lifts the burden of togetherness. Toward the goal of an affirmative biopolitical communitas, Esposito invites us to act from the commons, conceive objects as subjects, and make a politics of life, not over life. This affirmative biopolitics embraces nonhuman vitality and cosmopolitical diplomacy for multispecies life. It necessitates tough decisions, speciesist exclusions, and other triages. For as animal studies scholar Cary Wolfe states, an affirmative biopolitics cannot "simply embrace 'life' in all its undifferentiated singularity" (2012, 104). Rather, an affirmative biopolitics requires selecting certain organisms on which to focus the machinery of technics and governance. Stifled by technological failure, civil inaction, and elemental disturbance, this will not be completely successful or transparently fair—but choose we must. Having entered the compromise of blue governmentality, we will have picked one life for another—these seals and whales whose lives we story, and those we do not; these sharks in these waters we protect, those outside we do not. Blue governmentality is a compromise with life. It is processive, paradoxical, never complete, and rarely ideal. Most efforts fail, some may succeed. Try we must.

OCEAN/CULTURES

Blue governmentality aspires to ocean/cultures, a differentiation of marine from human life. Because blue governmentality is never complete, such ocean/cultures have yet to be. Their outlines, however, emerge in the achievements detailed in chapters 3, 4, and 7, in which activists use elemental technologies such as drones, ships, helicopters, flares, and water cannons to defend—and individuate—marine nature from human cultures. This activism is backed by conservationists and kindred scholars who argue that there is a complex entanglement of natures and cultures that should remain distinct in practice and theory. For them, large intact ecosystems must be protected so that the web of life might continue. They believe that the current sixth extinction is a grave existential crisis that must be immediately addressed with radical interventions. Technologies, policing, surveillance—nothing is off limits toward the goal of preserving a nature distinct from culture.

Whales are not harpooning themselves; sharks are not thrusting their bodies into nets or donating their fins for soup; terns are not dropping drones on their nests; nor are coral polyps intentionally dissolving their calcium structures. Humans are doing this to that. The limits of entanglement as an effective theory of new materialism are found where the agency of humans meets the mortal existence of nonhumans (Barad 2007). Everything appears as vital, vibrant, and interwoven matter (Bennett 2010)—until it is not. Extinction, death, and its certainties and entropies expose the poverty of entanglement theory and natureculturism.

Physical and epistemic intimacies between scientists and activists, and the people, fish, and animals they care for, do not synthesize these diverse beings into hybridities. Rather, tense differences, "intermittent attachments, and partial connections" characterize this "alongsidedness" (Latimer and Miele 2013, 15). Drone oceanography and activism forge intimate intra-actions that both enhance proximity (and their felt experience) and establish boundaries of difference. This pursuit follows geographer David Harvey, who invites us to "identify a restricted number of very general underlying processes which simultaneously *unify and differentiate* the phenomena we see in the world around us" (1996, 58). In eliminating the dichotomy of nature/culture, natureculturists tend toward monism, vitalism, and hybridization. Everything is everything else—connected, alive, and with alien agency. Avoiding this lumping requires scrutiny of how things both come together and fall apart. In their schema of "convivial conservation," conservation theorists Bram Büscher and Robert Fletcher appreciate the complementarity and modularity of parts and wholes: "we need to see both the integration and the separation" (2020, 130). Nature and culture are parallel, but we need to know the consistencies of each, the ways they influence and overwhelm each other, and the historical trajectories that surface from their interaction. A helix of separate yet intertwined strands. Oil and water in a liquid state. Alongsidedness graphically depicts the dynamic boundaries of nature that conservationists would like to see established and monitored by blue governmentality.

Geographer Jamie Lorimer (2015) rejects natureculturism. He is pragmatic about the role of conservation technologies in governing wild biopower. Instead of natureculture, Lorimer offers "multiple natures" to describe the "forms of natural knowledge—not all of which are scientific or even human—informing a myriad of discordant ways of living with the world" (2015, 2). "Multiple natures" are contingent configurations of convivial and cosmopolitical distinction and biopolitical asymmetry, requiring nurturing and sacrifice.

With pugnacious animality, elemental perturbations, and relative technicity's fallibility, blue governmentality is a necessity for convivial conservation. Lorimer chants, "Wildlife needs conservation. It needs science. It needs technology. It needs administration. It needs politics" (2015, 189). Environmental historian Andreas Malm agrees; we need "full alignment with cutting-edge science" (2018, 132). These are not naive, technoliberal sentiments of scholars who hope technoscientific gimmicks can overcome the contradictions of market expansion (Fish 2017). This is neither technosolutionism nor political solutionism, but an integration with the affirmative stance of nature realism. With the technologies and the politics of blue governmentality, we may preserve some nonhuman seas.

Modern humans are dependent on the sea. The ocean kept our young species alive during an early cataclysmic drought. Ill adapted to the colder and drier climes that arose between 123,000 and 195,000 years ago, the population of early *Homo sapiens* was in decline. As few as several hundred clung to existence in caves in South Africa, collecting shellfish and capturing shore fish (Marean 2012; Henshilwood et al. 2001). Their progeny made their way out of Africa and into Southeast Asia by land and continued to exploit marine resources. Traveling far from shore, by 50,000 years ago, modern humans were capable of long-distance sea travel. By 42,000 years ago humans were deep-sea fishing for tuna in Timor-Leste (*Thunnini*) (O'Connor, Rintaro, and Clarkson 2011). Increasingly, archaeologists believe that the peopling of North America from Eurasia occurred via the sea—not an interior ice-free corridor as was long thought (Wade 2017). The twenty-nine human footprints in beach sands dating to 13,000 years ago found on Calvert Island, British Columbia, are evocative evidence for humans' coastal connection (McLaren et al. 2018). Today, about 17 percent of the world's per capita consumption of animal protein comes from the ocean (Bennett et al. 2018). But with better management and shifts to protein sources that reproduce more quickly and sustainably, such as bivalves, oysters, mussels, and so on, the caloric bounty from the ocean could increase (Probyn 2016; Schubel and Thompson 2019). Economically, we are dependent on the ocean, generating $2.5 trillion annually for global economies (Ghosh 2020). We eat from the sea and breathe it in. The oxygen released as exhaust from photosynthesizing ocean plankton—plants, bacteria, and algae—contributes 50–80 percent of that gas in the atmosphere (NOAA n.d.). Phytoplankton requires the nitrogen distributed by whales, seals, and seabirds as they feed, dive, surface, defecate, die, and sink (Roman and McCarthy 2010). All of us are from the sea.

Historically, the oceans have mediated trade, warfare, colonialism, slavery, and primitive accumulation with their buoyancy and linear vectors for travel (Abulafia 2019). Today, with improved vessel construction, humans can spend more time—and extract more wealth from—the sea, its surface, and surrounding air. Despite not having gills or fins, we are becoming more capable in and around the sea. And yet it remains a volumetric and elemental space that prohibits our long-term habitation. The relative technicities of drones, ships, and airplanes provide temporary visits to the air or sea, but we are land mammals with oceanic livelihoods. Human bodies evolved for mobility in land-based gravity, and our technologies for oceanic exploration are relegated to traveling alongside and over, instead of living within or atop the seas for long. For these reasons, most humans have a sparser history and set of practices for the ocean than we do on land. Humans have been unable to incorporate the sea into our domestic and resource practices to the same extent we have incorporated land-based activities; in this way, the oceans remain an other to human culture.

This is a relative otherness, however, as much nature features some anthropogenic chemistry. As human presence increases, so too will oceancultures, requiring the application of conservation technologies and the blue governmentality of ocean/cultures to protect biodiversity. To coexist with oceans and cultures alongside each other as ocean/cultures, we need oceanic thinking that flows with the materiality of the air and turbulence of the sea. The ocean's elementality and the atmosphere's mobility make possible forms of control unlike that which the hard ground allows.

Natureculture is thought from the firmament. Oceanic thinking challenges this terrestrial-centrism. In the ocean, more fluid concepts emerge. Saturation, turbulence, erosion, turbidity, corrosion, and other ocean-centric concepts contribute to a more nuanced and situated understanding of the relationship between oceans and cultures (Jue and Ruiz 2021; Lehman, Steinberg, and Johnson 2021). This book offers blue governmentality and ocean/cultures as additional ways of conceptualizing ocean ontology and conservation. The hybridization of natureculture will come into greater resolution through the ocean's absorption of microplastics, dissolved acid, atmospheric warming, and other anthropogenic factors. Increasing industrial fishing, both targeting large keystone species and working down the trophic cascades to smaller fish, will continue to modify the stability of ecological relationships. This is more reason for separating oceans from cultures.

Elementality, technicity, and political processes—these and other actants configure blue governmentality. The conservation potential of blue governmentality rests on the translation of surveillance data into policy and prosecution. Countless images and samples have been collected, as has evidence of egregious violations of fishing regulations. Technology's influence on conservation will be established through more traditional modes of human governance—laws and regulations. But neither drone-derived scientific nor legal evidence has substantially assisted the protection of endangered marine habitats or species. This is a simple observation, and one I hope will be proven wrong in the future. Without this, the drone will continue to be a revolutionary data-collecting tool with limited impacts on prospering.

In this book's cases, the criminals largely avoid punitive damage. Conservation drones are not the pernicious, invasive, and dominating instruments that cultural geographers and media studies scholars fear. Drone power dissipates before entering the courtrooms in Mexico, Timor-Leste, and the United States. Experiments throughout the world are focused on connecting drone data to the prosecution of aquatic crime—catching salmon poachers in Scotland (Gerrard 2021), monitoring illegal activity in marine parks in Australia, finding illegal fishing in the Seychelles (United Nations Environmental Program 2018), and fitting albatrosses with sensors capable of identifying illegal fishing around New Zealand—"They're like drones, only intelligent," noted one of the participating scientists (Roy 2020).

The drone's technological achievements are modest acts of nature surveillance. Not preserving marine lives, these drones are not yet conservation drones. This is not a natureculture wherein marine populations and their threats are accurately captured by technologies before being integrated into comprehensive management and enforcement strategies. Marine biopower has not given in to technosolution managerialism. Despite increasing surveillance by ships, satellites, drones, smart buoys, sinking sensors, and instrumented animals, the ocean has not transformed into a sea that is cybernetically controlled. For blue governmentality to convincingly emerge, absent components need integration: data analysis and interpretation as well as management and enforcement in line with the biopolitical stack.

For many scholars, biopolitics and surveillance are universally negative —no affirmative biopolitics can exist, only necropolitics. This cynicism could be tempered by the possibility that extinction looms if biopolitical surveillance and governance do not propagate. Unless there are significant changes in human fish consumption and in ocean acidification and pollution, and unless technologies for policing illegal and unregulated fishing are deployed in significant numbers in conjunction with punishment for violations, oceanic biomass will continue to drop.

And still, poachers escape. Drones are technologies with partial efficacy. Constrained by the elements, limited by the relative applications of their technicity, drones are not everywhere, nor do they see everything. Despite these limitations, the uptake of drones in conservation and activism will continue to rise. The drone's expansion into the adjacent possible—delivery and inspection industries, search and rescue, and policing and humanitarianism—is evidence that originary technicity's evolutionary tendency is to find and fill niches where speed, mobility, and vision are privileged. As drone use increases in conservation, and drone data becomes more persuasive as evidence of criminality in prosecutions, drones will begin to contribute more substantially to the tightening of ocean/cultures. As this happens, the consequentiality of blue governmentality for marine flourishing will sharpen. This ocean/culture to emerge, replete with surveillance drones with more linear legal impacts, will not be the autonomous ocean that marine activists struggle to protect. And yet, technological foibles, general entropy, and elemental contingencies will foil this speculative ocean/culture of the future. From the perspective of nature realism, even this creeping possibility of natureculturism is preferable to extinction.

Scientific hypotheses are generated through data collection and analysis. Publics and politicians debate in networked yet filtered media. Rationality, economic pragmatism, and emotional responses compete. The protocols of conservation law, regulations, policing, vigilante justice, and marine parks are established, challenged, defended, or negated. Technologies of science, politics, and activism are deployed—or not. When they are, the elements, human pilots and regulations, and animal agency conspire with a limited impact on the protection of marine ecologies. Never final and often affirmative, much wildness will escape this blue governmentality. This fugitive ocean is as it should be—a precious medium for multispecies vivacity and a mystery that remains out of control.

Abulafia, David. 2019. *The Boundless Sea: A Human History of the Oceans*. Oxford: Oxford University Press.

Adams, William. 2017. "Geographies of Conservation II: Technology, Surveillance and Conservation by Algorithm." *Progress in Human Geography* 43, no. 2 (November): 337–50.

Adey, P. 2015. "Air's Affinities: Geopolitics, Chemical Affect and the Force of the Elemental." *Dialogues in Human Geography* 5, no. 1 (March): 54–75.

Alaimo, Stacy. 2012. "States of Suspension: Trans-corporeality at Sea." *ISLE: Interdisciplinary Studies in Literature and Environment* 19, no. 3 (December): 476–93.

Alaimo, Stacy. 2013. "Jellyfish Science, Jellyfish Aesthetics: Posthuman Reconfigurations of the Sensible." In *Thinking with Water*, edited by C. Chen, J. MacLeod, and A. Neimanis, 139–64. Montreal: McGill–Queen's University Press. http://www.jstor.org/stable/j.ctt32b7pn.17.

Alexiadou, Paraskevi, Ilias Foskolos, and Alexandros Frantzis. 2019. "Ingestion of Macroplastics by Odontocetes of the Greek Seas, Eastern Mediterranean: Often Deadly!" *Marine Pollution Bulletin* 146 (September) 67–75. https://doi.org/10.1016/j.marpolbul.2019.05.055.

Ampou, Elvan Eghbert, Sylvain Oillon, Corina Iovan, and Serge Andréfouët. 2018. "Change Detection of Bunaken Island Coral Reefs Using 15 Years of Very High-Resolution Satellite Images: A Kaleidoscope of Habitat Trajectories." *Marine Pollution Bulletin* 131 (June): 83–95. https://doi.org/10.1016/j.marpolbul.2017.10.067.

Anderson, Chris. 2017. "Drones Go to Work." *Harvard Business Review*, May 16, 2017. https://hbr.org/2017/05/drones-go-to-work.

Anderson, Jon. 2012. "Relational Places: The Surfed Wave as Assemblage and Convergence." *Environment and Planning D: Society and Space* 30 (4): 570–87.

Andrews, Crispin. 2014. "UAVs in the Wild." *Engineering and Technology* 9 (7): 33–35. https://doi.org/10.1049/et.2014.0705.

Angier, Natalie. 2010. "Save a Whale, Save a Soul, Goes the Cry." *New York Times*, June 26, 2010. https://www.nytimes.com/2010/06/27/weekinreview/27angier.html.

Apprill, A., C. A. Miller, M. J. Moore, D. W. Durban, H. Fearnbach, and L. G. Barrett-Lennard. 2017. "Extensive Core Microbiome in Drone-Captured

Whale Blow Supports a Framework for Health Monitoring." *mSystems* 2 (5): e00119–17.

Archibald, Jo-ann Q'um Q'um Xiiem, Jenny Bol Jun Lee-Morgan, and Jason De Santolo. 2019. *Decolonizing Research: Indigenous Storywork as Methodology*. London: Zed.

Asker, Chloe. 2020. "(Mc)Mindfulness and the Politics of Breathing." *Life of Breath* (blog), May 12, 2020. https://lifeofbreath.webspace.durham.ac.uk /mcmindfulness-and-the-politics-of-breath/.

Atkinson, Shannon, Andrew Rogan, C. Scott Baker, Ralf Dagdag, Matthew Redlinger, Jennifer Polinski, Jorge Urban, et al. 2021. "Genetic, Endocrine, and Microbiological Assessments of Blue, Humpback and Killer Whale Health Using Unoccupied Aerial Systems." *Wildlife Society Bulletin* 45, no. 4 (December): 654–69.

Australian Government. N.d. "Crown-of-Thorns Starfish Control Program." Great Barrier Reef Marine Park Authority. Accessed March 2, 2022. https://www .gbrmpa.gov.au/our-work/our-programs-and-projects/crown-of-thorns -starfish-management/crown-of-thorns-starfish-control-program.

Australian Shark Incident Database. N.d. Taronga Conservation Society Australia. Accessed March 2, 2022. https://taronga.org.au/conservation-and-science /australian-shark-incident-database.

Avron, Lisa. 2017. "'Governmentalities' of Conservation Science at the Advent of Drones: Situating an Emerging Technology." *Information and Culture: A Journal of History* 52 (3): 362–83. https://doi.org/10.1353/lac.2017.0014.

Badiou, Alain. 2005. *Being and Event*. New York: Continuum.

Bailey, Kevin. 2011. "An Empty Donut Hole: The Great Collapse of a North American Fishery." *Ecology and Society* 16 (2): 28.

Bailey, K., T. Quinn, P. Bentzen, and W. S. Grant. 2000. "Population Structure and Dynamics of Walleye Pollock, *Theragra chalcogramma*." *Advances in Marine Biology* 37:179–255.

Ballestero, Andrea. 2019. "Touching with Light, or, How Texture Recasts the Sensing of Subterranean Water Worlds." *Science, Technology, and Human Values* 44 (4).

Balmford, Andrew. 2012. *Wild Hope: On the Front Lines of Conservation Success*. Chicago: University of Chicago Press.

Banse, Tom. 2016. "Hobby Drones over Orca Whales: Legally Murky and a Potentially Bad Mix." Northwest News Network, May 13, 2016. https://www .nwnewsnetwork.org/crime-law-and-justice/2016-05-13/hobby-drones-over -orca-whales-legally-murky-and-a-potentially-bad-mix.

Barad, Karen. 2003. "Posthumanist Performativity: Toward an Understanding of How Matter Comes to Matter." *Signs: Journal of Women in Culture and Society* 28 (3): 801–31.

Barad, Karen. 2007. *Meeting the Universe Halfway: Quantum Physics and the Entanglement of Matter and Meaning*. Durham, NC: Duke University Press.

Barker, Anne. 2018. "'Something Is Not Right': How $US100,000 Ensured a Million-Dollar Illegal Catch Was Forgotten in East Timor." ABC News, June

29, 2018. https://www.abc.net.au/news/2018-06-30/million-dollar-illegal
-catch-forgotten/9925890.

Baudrillard, Jean. 1983. *Simulations*. Translated by P. Foss, P. Beitchman, and
P. Patton. Los Angeles: Semiotext(e).

Bear, Christopher, and Sally Eden. 2011. "Thinking Like a Fish? Engaging with
Nonhuman Difference through Recreational Angling." *Environment and Planning D: Society and Space* 29, no. 2 (January): 336–52.

Beck, Ulrich. 1992. *Risk Society: Towards a New Modernity*. New Delhi: Sage.

Beldo, Les. 2019. *Contesting Leviathan: Activists, Hunters, and State Power in the
Makah Whaling Conflict*. Chicago: University of Chicago Press.

Bennett, Abigail, Patil Pawan, Kristin Kleisner, Doug Rader, John Virdin, and
Xavier Basurto. 2018. *Contribution of Fisheries to Food and Nutrition Security:
Current Knowledge, Policy, and Research*. Nicholas Institute for Environmental
Policy Solutions report 18-02. Durham, NC: Duke University. https://
nicholasinstitute.duke.edu/sites/default/files/publications/contribution
_of_fisheries_to_food_and_nutrition_security_0.pdf.

Bennett, Jane. 2010. *Vibrant Matter: A Political Ecology of Things*. Durham, NC:
Duke University Press.

Bennett, Nathan James, and Philip Dearden. 2014. "Why Local People Do Not
Support Conservation: Community Perceptions of Marine Protected Area
Livelihood Impacts, Governance and Management in Thailand." *Marine Policy*
44 (February): 107–16.

Benson, Etienne. 2010. *Wired Wilderness: Technologies of Tracking and the Making of
Modern Wildlife*. Baltimore: Johns Hopkins University Press.

Benson, Etienne. 2011. "Autonomous Biological Sensor Platforms." *Cabinet*
41:75–78.

Benson, Etienne. 2013. "Demarcating Wilderness and Disciplining Wildlife: Radio-
tracking Large Carnivores in Yellowstone and Chitwan National Parks." In
Civilizing Nature: National Parks in Global Historical Perspective, edited by Bern-
hard Gißibl, Sabine Höhler, and Patrick Kupper, 173–88. New York: Berghahn.

Blair, James J. A. 2022. "Tracking Penguins, Sensing Petroleum: 'Data Gaps' and
the Politics of Marine Ecology in the South Atlantic." *Environment and Planning E: Nature and Space* 5 (1): 60–80.

Boas, Franz. 1940. *Race, Language, and Culture*. New York: Macmillan.

Boehlert, G. W., D. P. Costa, D. E. Crocker, P. Green, T. O'Brien, S. Levitus, and
B. J. Le Boeuf. 2001. "Autonomous Pinniped Environmental Samplers: Using
Instrumented Animals as Oceanographic Data Collectors." *Journal of Atmo-
spheric and Oceanic Technology* 18 (11): 1882–93.

Boler, Megan. 2008. "Toward Open and Dense Networks: An Interview with
Geert Lovink." In *Digital Media and Democracy: Tactics in Hard Times*, edited by
Megan Boler, 123–36. Cambridge, MA: MIT Press.

Bolin, Jessica, David Schoeman, Carme Pizà-Roca, and Kylie Scales. 2020. "A Current Affair: Entanglement of Humpback Whales in Coastal Shark-Control
Nets." *Remote Sensing in Ecology and Conservation* 6, no. 2 (November): 119–28.

Bolla, G. M., M. Casagrande, A. Comazzetto, R. Dal Moro, M. Destro, E. Fantin, G. Colombatti, et al. 2018. "ARIA: Air Pollutants Monitoring Using UAVs." In *2018 5th IEEE International Workshop on Metrology for AeroSpace (MetroAeroSpace)*, 225–29. IEEE. http://doi.org/10.1109/MetroAeroSpace.2018.8453584.

Bonaccorso, E., N. Ordóñez-Garza, D. A. Pazmiño, A. Hearn, D. Páez-Rosas, S. Cruz, J. Pablo Muñoz-Pérez, et al. 2021. "International Fisheries Threaten Globally Endangered Sharks in the Eastern Tropical Pacific Ocean: The Case of the *Fu Yuan Yu Leng 999* Reefer Vessel Seized within the Galápagos Marine Reserve." *Scientific Reports* 11, art. no. 14959 (July). https://doi.org/10.1038/s41598-021-94126-3.

Bookchin, Murray. 1971. *Post-scarcity Anarchism*. Berkeley, CA: Ramparts. Reprint, Montreal: Black Rose, 1986; and San Francisco: AK, 2004.

Borell, Andre, dir. 2021. *Envoy: Shark Cull*. Hype Project, Australia.

Bossart, G. D. 2011. "Marine Mammals as Sentinel Species for Oceans and Human Health." *Veterinary Pathology* 48 (3): 676–90. https://doi.org/10.1177/0300985810388525.

Boucquey, Noëlle, Luke Fairbanks, Kevin St. Martin, Lisa M. Campbell, and Bonnie McCay. 2016. "The Ontological Politics of Marine Spatial Planning: Assembling the Ocean and Shaping the Capacities of 'Community' and 'Environment.'" *Geoforum* 75, no. 2 (October): 1–11.

Boucquey, Noëlle, Kevin St. Martin, Luke Fairbanks, Lisa M. Campbell, and Sarah Wise. 2019. "Ocean Data Portals: Performing a New Infrastructure for Ocean Governance." *Environment and Planning D: Society and Space* 37 (3): 484–503.

Bradshaw, Corey, Paul R. Ehrlich, Andrew Beattie, Gerardo Ceballos, Eileen Crist, Joan Diamond, Rodolfo Dirzo, et al. 2021. "Underestimating the Challenges of Avoiding a Ghastly Future." *Frontiers in Conservation Science* 1, art. no. 615419. https://doi.org/10.3389/fcosc.2020.615419.

Braidotti, Rosi. 2013. *The Posthuman*. Hoboken, NJ: Wiley.

Braidotti, Rosi. 2019a. *Posthuman Knowledge*. Cambridge: Polity.

Braidotti, Rosi. 2019b. "A Theoretical Framework for the Critical Posthumanities." *Theory, Culture and Society* 36 (6): 31–61.

Braidotti, R. 2020. "'We' May Be in This Together, but We Are Not All Human and We Are Not One and the Same." *Ecocene: Cappadocia Journal of Environmental Humanities* 1 (1): 1–31.

Bratton, Benjamin. 2021. *The Revenge of the Real: Politics for a Post-pandemic World*. London: Verso.

Brautigan, Richard. 1968. *All Watched Over by Machines of Loving Grace*. San Francisco: Communications Company. http://www.brautigan.net/machines.html.

Braverman, Irus. 2014. "Governing the Wild: Databases, Algorithms, and Population Models as Biopolitics." *Surveillance and Society* 12 (1): 15–37.

Braverman, Irus. 2020. "Robotic Life in the Deep Sea." In *Blue Legalities: The Life and Laws of the Sea*, edited by Irus Braverman and Elizabeth R. Johnson, 147–64. Durham, NC: Duke University Press.

Braverman, Irus, and Elizabeth R. Johnson, eds. 2020. *Blue Legalities: The Life and Laws of the Sea*. Durham, NC: Duke University Press.

Brighenti, A. 2007. "Visibility: A Category for the Social Sciences." *Current Sociology* 55 (3): 323–42.

Brockington, Dan. 2002. *Fortress Conservation: The Preservation of the Mkomazi Game Reserve, Tanzania*. Oxford: James Currey.

Burnett, D. Graham. 2012. *The Sounding of the Whale: Science and Cetaceans in the Twentieth Century*. Chicago: University of Chicago Press.

Büscher, Bram. 2016. "'Rhino Poaching Is Out of Control!' Violence, Race and the Politics of Hysteria in Online Conservation." *Environment and Planning A: Economy and Space* 48 (5): 979–98.

Büscher, Bram, and Robert Fletcher. 2020. *The Conservation Revolution: Radical Ideas for Saving Nature beyond the Anthropocene*. London: Verso.

Büscher, Bram, and Maano Ramutsindela. 2016. "Green Violence: Rhino Poaching and the War to Save Southern Africa's Peace Parks." *African Affairs* 115 (458): 1–22.

Butcher, Paul A., Toby P. Piddocke, Andrew P. Colefax, Brent Hoade, Victor M. Peddemors, Lauren Borg, and Brian R. Cullis. 2019. "Beach Safety: Can Drones Provide a Platform for Sighting Sharks?" *Wildlife Research* 46, no. 8 (December): 701–12.

Butler, Judith. 1997. *The Psychic Life of Power: Theories in Subjection*. Stanford, CA: Stanford University Press.

Calvillo, Nerea, and Emmas Garnett. 2019. "Data Intimacies: Building Infrastructures for Intensified Embodied Encounters with Air Pollution." *Sociological Review* 67 (2): 340–56.

Canguilhem, Georges. 1965. *La Connaissance de la vie*. Paris: Vrin.

Canguilhem, Georges, and John Savage (translator). 2001. "The Living and Its Milieu." *Grey Room*, no. 3 (Spring), 7–31.

Cansdale, Dominic, and Sarah Cummings. 2020. "Humpback Whale off Burleigh Heads Freed from Shark Nets by Man in Tinny." ABC News, May 18. https://www.abc.net.au/news/2020-05-19/whale-caught-in-gold-coast-shark-nets/12261980.

Carbone, Christopher. 2018. "Hero Whale Saves Snorkeler from Tiger Shark in the Pacific Ocean." Fox News, January 8, 2018. https://www.foxnews.com/science/hero-whale-saves-snorkeler-from-tiger-shark-in-the-pacific-ocean.

Castells, Manuel. 1996. *The Rise of the Network Society*. Vol. 1, *The Information Age: Economy, Society, and Culture*. Malden, MA: Blackwell.

Castree, Noel. 2014. *Making Sense of Nature: Knowledge, Politics and Democracy*. London: Routledge.

Ceballos, G., P. R. Ehrlich, and P. H. Raven. 2020. "Vertebrates on the Brink as Indicators of Biological Annihilation and the Sixth Mass Extinction." *Proceedings of the National Academy of Sciences of the United States of America* 117 (24): 13596–602.

Chamayou, Gregoire. 2015. *Drone Theory*. London: Penguin.

Chappell, Bill. 2012. "Japanese Whalers Lose Bid to Block U.S.-Based 'Sea Shepherd' Activists." *The Two-Way*, National Public Radio, February 16, 2012. https://www.npr.org/sections/thetwo-way/2012/02/16/147012056/japanese-whalers-lose-bid-to-block-u-s-based-sea-shepherd-activists.

Chapron, Guillaume, Petra Kaczensky, John Linnell, Manuela von Arx, Djuro Huber, Henrik Andrén, José Vicente López-Bao, et al. 2014. "Recovery of Large Carnivores in Europe's Modern Human-Dominated Landscapes." *Science* 346, no. 6216 (December): 1517–19. https://doi.org/10.1126/science.1257553.

Charrassin, Jean-Benoît, Young-Hyang Park, Yvon Le Maho, and Charles-André Bost. 2002. "Penguins as Oceanographers Unravel Hidden Mechanisms of Marine Productivity." *Ecology Letters* 5, no. 3 (May): 317–19. https://doi.org/10.1046/j.1461-0248.2002.00341.x.

Choy, T., and J. Zee. 2015. "Condition—Suspension." *Cultural Anthropology* 30, no. 2 (May): 210–23. https://doi.org/10.14506/ca30.2.04.

Christiansen, F., A. M. Dujon, K. R. Sprogis, J. P. Y. Arnould, and L. Bejder. 2016. "Noninvasive Unmanned Aerial Vehicle Provides Estimates of the Energetic Cost of Reproduction in Humpback Whales." *Ecosphere* 7, no. 10 (October): e01468. https://doi.org/10.1002/ecs2.1468.

Chun, Wendy Hui Kyong. 2021. *Discriminating Data: Correlation, Neighborhoods, and the New Politics of Recognition*. Cambridge, MA: MIT Press.

Cohen, Jeffrey Jerome, and Lowell Duckert. 2015. "Introduction: Eleven Principles of the Elements." In *Elemental Ecocriticism: Thinking with Air, Water, and Fire*, edited by Jeffrey Jerome Cohen and Lowell Duckert, 1–26. Minneapolis: University of Minnesota Press.

Cooper, Luke. 2020. "Gold Coast Whale Rescuer 'Django' to Donate Fine Fund to Conservation Group." 9 News, May 21, 2020. https://www.9news.com.au/national/gold-coast-whale-rescuer-django-escapes-fine-but-receives-official-warning-over-actions-qld-news/e843961b-9367-442c-b8c3-c5669782b58b.

Corrigan, C. E., G. C. Roberts, M. V. Ramana, D. Kim, and V. Ramanathan. 2008. "Capturing Vertical Profiles of Aerosols and Black Carbon over the Indian Ocean Using Autonomous Unmanned Aerial Vehicles." *Atmospheric Chemistry and Physics* 8, no. 3 (February): 737–47. https://doi.org/10.5194/acp-8-737-2008.

Coxworth, Ben. 2019. "Amphibious Drone Designed to Warn of Wonky Water." *New Atlas*, October 16, 2019. https://newatlas.com/drones/pelican-water-sampling-drone/.

Crandall, Jordan. 2011. "Ontologies of the Wayward Drone: A Salvage Operation." *CTheory*, November 2, 2011. https://journals.uvic.ca/index.php/ctheory/article/view/14955.

Crilly, Rob. 2018. "Orca Mother Keeps Her Dead Calf Afloat for Two Days in Extraordinary Display of Grief." *Telegraph*, July 27, 2018. https://www.telegraph.co.uk/news/2018/07/27/orca-mother-keeps-dead-calf-afloat-extraordinary-display-grief/.

CrimethInc. 2020. "A Demonstrator's Guide to Gas Masks and Goggles: Everything You Need to Know to Protect Your Eyes and Lungs from Gas and Projectiles." CrimethInc., September 2, 2020. https://crimethinc.com/2020/09/02/a-demonstrators-guide-to-gas-masks-and-goggles-everything-you-need-to-know-to-protect-your-eyes-and-lungs-from-gas-and-projectiles.

Crist, Eileen. 2013. "On the Poverty of Our Nomenclature." *Environmental Humanities* 3 (1): 129–47.

Crist, Eileen. 2015. "I Walk in the World to Love It." In *Protecting the Wild: Parks and Wilderness, the Foundation for Conservation*, edited by George Wuerthner and Eileen Crist, 82–95. Washington, DC: Island.

Cronon, William. 1996. "The Trouble with Wilderness; or, Getting Back to the Wrong Nature." *Environmental History* 1, no. 1 (January): 7–28.

Dartnell, Jo, and Michelle Ramsay. 2005. *One Stop Doc: Respiratory System*. London: Hodder Arnold.

Dean, Jodi. 2005. "Communicative Capitalism: Circulation and the Foreclosure of Politics." *Cultural Politics* 1 (1): 51–74.

Dean, Mitchell, and Daniel Zamora. 2021. *The Last Man Takes LSD: Foucault and the End of Revolution*. London: Verso.

Debord, Guy. 1994. *The Society of the Spectacle*. New York: Zone.

Dedrone. 2019. "Worldwide Drone Incidents: All Drone Incidents Documented by the Press Worldwide." https://www.dedrone.com/resources/incidents-new/all.

Deleuze, Gilles. (1970) 1988. *Spinoza: Practical Philosophy*. San Francisco: City Lights.

Deleuze, Gilles. 1992. "Postscript on the Societies of Control." *October* 59 (Winter): 3–7.

Deleuze, Gilles, and Félix Guattari. (1972) 1977. *Anti-Oedipus: Capitalism and Schizophrenia*. Translated by Robert Hurley and Mark Seem. New York: Viking Penguin.

Deloria, Vine, Jr. 2001. "American Indian Metaphysics." In *Power and Place: Indian Education in America*, edited by Vine Deloria Jr. and Daniel R. Wildcat, 1–6. Golden, CO: Fulcrum.

De Robertis, A., N. Lawrence-Slavas, R. Jenkins, I. Wangen, C. W. Mordy, C. Meinig, M. Levine, et al. 2019. "Long-Term Measurements of Fish Backscatter from Saildrone Unmanned Surface Vehicles and Comparison with Observations from a Noise-Reduced Research Vessel." *ICES Journal of Marine Science* 76 (7): 2459–70.

Derrida, Jacques. (1967) 1998. *Of Grammatology*. Baltimore: Johns Hopkins University Press.

Descola, Philippe. 2012. "Beyond Nature and Culture." *HAU: Journal of Ethnographic Theory* 2, no. 1 (Spring 2012): 447–71. 10.14318/hau2.1.021.

Di Stefano, G., G. Romeo, A. Mazzini, A. Iarocci, S. Hadi, and S. Pelphrey. 2018. "The Lusi Drone: A Multidisciplinary Tool to Access Extreme Environments." *Marine and Petroleum Geology* 90 (February): 26–37.

Djudjic, Dunja. 2019. "This Algorithm Removes Water from Underwater Photos, and They Look Incredible." *Photography*, November 14, 2019. https://www.diyphotography.net/this-algorithm-removes-water-from-underwater-photos-and-they-look-incredible/.

Dobbs, Kirstin. 2007. "A Reef-Wide Framework for Managing Traditional Use of Marine Resources in the Great Barrier Reef Marine Park." Great Barrier Reef Marine Park Authority, Australian Government. https://elibrary.gbrmpa.gov.au/jspui/handle/11017/408.

Dōgen, Eihei. 1995. *Moon in a Dewdrop: Writings of Zen Master Dōgen* [b. 1200]. Edited by Kazuaki Tanahashi. Berkeley, CA: North Point.

Dolman, Trish, dir. 2011. *EcoPirate: The Story of Paul Watson*. Screen Siren Pictures and Optic Nerve Films.

Donahue, Michelle. 2016. "Meet the Dogs Sniffing Out Whale Poop for Science." *Smithsonian Magazine*, February 8, 2016. https://www.smithsonianmag.com/science-nature/meet-dogs-sniffing-out-whale-poop-science-180958050/.

Donnison, Jon. 2014. "Australians Protest against Shark 'Cull.'" BBC News, January 4, 2014. https://www.bbc.com/news/av/world-asia-pacific-25603844.

Downing, John D. H. 2001. *Radical Media: Rebellious Communication and Social Movements*. Thousand Oaks, CA: SAGE.

Drury, J., L. Riek, and N. Rackliffe. 2006. "A Decomposition of UAV-Related Situation Awareness." In *Proceedings of the 1st ACM SIGCHI/SIGART Conference on Human-Robot Interaction*, 88–94. New York: Association for Computing Machinery.

Duffy, Rosaleen, Francis Massé, Emile Smidt, Esther Marijnen, Bram Büscher, Judith Verweijen, Maano Ramutsindela, et al. 2019. "Why We Must Question the Militarisation of Conservation." *Biological Conservation* 232 (April): 66–73.

Dunagan, Christopher. 2018. "Orca Bill Sinks in State Senate." King 5, February 18, 2018. https://www.king5.com/article/news/local/orca-bill-sinks-in-state-senate/281-520472928.

Earth First! 2005. "Sea Shepherd 'Spikes' Ocean Near Newfoundland." *Earth First!: The Radical Environmental Journal* 25, no. 5 (July–August): 14. https://www.environmentandsociety.org/sites/default/files/key_docs/ef_25_5_3.pdf.

Eilstrup-Sangiovanni, Mette, and Teale N. Phelps Bondaroff. 2014. "From Advocacy to Confrontation: Direct Enforcement by Environmental NGOs." *International Studies Quarterly* 58 (2): 348–61.

Elhacham, Emily, Liad Ben-Uri, Jonathan Grozovski, Yinon M. Bar-On, and Ron Milo. 2020. "Global Human-Made Mass Exceeds All Living Biomass." *Nature* 588, no. 7838 (December): 442–44. https://doi.org/10.1038/s41586-020-3010-5.

Engelmann, S., and D. McCormack. 2021. "Elemental Worlds: Specificities, Exposures, Alchemies." *Progress in Human Geography* 45, no. 6 (December): 1419–39.

Esposito, Roberto. 2013. "Community, Immunity, Biopolitics." *Angelaki: Journal of the Theoretical Humanities* 18, no. 3 (November): 83–90.

Fairbanks, Luke, Noëlle Boucquey, Lisa Campbell, and Sarah Wise. 2019. "Remaking Oceans Governance: Critical Perspectives on Marine Spatial Planning." *Environment and Society* 10 (1): 122–40.

Fanner, David. 2019. "Shark Spotting: How a Drone App Is Helping to Keep Sydney's Surfers Safe in the Water." *Guardian*, January 22, 2019. https://www.theguardian.com/environment/video/2019/jan/22/drone-surveillance-spots-sharks-off-sydneys-beaches-its-a-big-one-video.

FAO. 2014. "The State of World Fisheries and Aquaculture: Opportunities and Challenges." Rome: Food and Agriculture Organization of the United Nations. https://www.fao.org/3/i3720e/i3720e.pdf.

Fedak, Michael. 2004. "Marine Animals as Platforms for Oceanographic Sampling: A 'Win/Win' Situation for Biology and Operational Oceanography." *Memoirs of the National Institute of Polar Research* 58:133–47.

Feigenbaum, Anna. 2014. "Resistant Matters: Tents, Tear Gas and the 'Other Media' of Occupy." *Communication and Critical/Cultural Studies* 11 (1): 15–24.

Feigenbaum, Anna. 2015. "From Cyborg Feminism to Drone Feminism: Remembering Women's Anti-nuclear Activisms." *Feminist Theory* 16 (3): 265–88.

Ferguson, M. C., R. P. Angliss, A. Kennedy, B. Lynch, A. Willoughby, V. Helker, A. A. Brower, et al. 2018. "Performance of Manned and Unmanned Aerial Surveys to Collect Visual Data and Imagery for Estimating Arctic Cetacean Density and Associated Uncertainty." *Journal of Unmanned Vehicle Systems* 6 (3): 128–54.

Firozi, Paulina. 2021. "Thousands of Eggs Abandoned after a Drone Scares Off Nesting Birds." *Washington Post*, June 7, 2021. https://www.washingtonpost.com/science/2021/06/07/drone-crash-abandoned-eggs/.

Fish, Adam. 2017. *Technoliberalism and the End of Participatory Culture in the United States*. London: Palgrave Macmillan.

Fish, Adam. 2019a. *Crash Theory*. YouTube video, May 29. https://www.youtube.com/watch?v=LEj8ECbBJe8.

Fish, Adam. 2019b. "Drones at the Edge of Naturecultures." *Media Fields Journal: Critical Explorations in Media and Space*, no. 14. http://mediafieldsjournal.squarespace.com/storage/issue-14/article-pdfs/1412-Fish-PDF.pdf

Fish, Adam. 2021. "Crash Theory: Entrapments of Conservation Drones and Endangered Megafauna." *Science, Technology, and Human Values* 46, no. 2 (March): 425–51. https://doi.org/10.1177/0162243920920356.

Fish, Adam. 2022a. "Blue Governmentality: Elemental Activism with Conservation Technologies on Plundered Seas." *Political Geography* 93, art. no. 102528 (March): n.p. https://doi.org/10.1016/j.polgeo.2021.102528.

Fish, Adam. 2022b. "Saildrones and Snotbots in the Blue Anthropocene: Sensing Technologies, Multispecies Intimacies, and Scientific Storying." Environment and Planning D: Society and Space 40, no. 5 (October): 862–80. https://doi.org/10.1177/02637758221126526.

Fish, Adam. 2023. "Technoliberalism and the Complementary Relationships between Humanitarian, Conservation, and Entrepreneurial Dronework in

Indonesia." *Convergence*. Published ahead of print, March 22, 2023. https://doi
.org/10.1177/13548565231161254.

Fish, Adam, and Luca Follis. 2016. "Gagged and Doxed: Hacktivism's Self-
Incrimination Complex." *International Journal of Communication* 10:3281–3300.

Fish, Adam, and Bradley L. Garrett. 2019. "Resurrection from Bunkers and Data
Centers." *Culture Machine* 18. https://culturemachine.net/wp-content
/uploads/2019/04/FISH.pdf.

Fish, Adam, Bradley Garrett, and Oliver Case. 2017a. "Drones Caught in the Net."
Imaginations: Journal of Cross-Cultural Image Studies 8 (2): 74–79. https://doi
.org/10.17742/IMAGE.LD.8.2.8.

Fish, Adam, Bradley Garrett, and Oliver Case. 2017b. "Extended Flight: The Emergence of Drone Sovereignty." *InVisible Culture: An Electronic Journal for Visual
Culture* 27:1–8.

Fish, Adam, and Michael Richardson. 2022. "Drone Power: Conservation,
Humanitarianism, Policing and War." *Theory, Culture and Society* 39 (3): 3–26.

Fish, Adam, and Ramesh Srinivasan. 2012. "Digital Labor Is the New Killer App."
New Media and Society 14 (1): 137–52.

Flitton, Daniel. 2017. "Economy of Scales: Depleted Stocks Force Asian Fishermen
into Australian Waters." *Sydney Morning Herald*, February 24, 2017. https://
www.smh.com.au/national/economy-of-scales-depleted-stocks-force-asian
-fishermen-into-australian-waters-20170224-guklc2.html.

Forssman, Natalie A. 2017. "Staging the Animal Oceanographer: An Ethnography
of Seals and Their Scientists." PhD diss., University of California, San Diego.

Foster, Arthur Ernest. 1922. "The First Coogee Shark Net, December 1922." *Dictio-
nary of Sydney*. https://dictionaryofsydney.org/media/5245.

Foucault, Michel. 1977. *Discipline and Punish: The Birth of the Prison*. New York:
Vintage.

Foucault, Michel. (1984) 1990. *The History of Sexuality*. Vol. 3, *The Care of the Self*.
London: Penguin.

Foucault, Michel. 1996. "Sex, Power and the Politics of Identity." In *Foucault Live:
Collected Interviews, 1961–1984*, edited by Sylvère Lotringer, 384. Cambridge,
MA: MIT Press.

Foucault, Michel. 1998. "Theatrum Philosophicum." In *Aesthetics, Method, and
Epistemology: Essential Works of Foucault 1954–1984*, vol. 2, edited by James D.
Faubion, 343. New York: New Press.

Foucault, Michel. 2003. *"Society Must Be Defended": Lectures at the Collège de
France, 1975–76*. New York: Picador.

Foucault, Michel. 2008. *The Birth of Biopolitics: Lectures at the Collège de France,
1978–1979*. Edited by Michel Senellart, François Ewald, and Alessandro Fon-
tana. New York: Picador.

Fox, Alex. 2019. "When Green Monkeys Spy a Drone, They Use Their Cousins' Cry
for 'Eagle.'" *Science*, May 27. https://doi.org/10.1126/science.aay1612.

Gabrys, Jennifer. 2016. *Program Earth*. Minneapolis: University of Minnesota
Press.

Gabrys, Jennifer, and Helen Pritchard. 2018. "Sensing Practices." In *Posthuman Glossary*, edited by Rosi Braidotti and Maria Hlavajova, 394–96. London: Bloomsbury.

Gaines, Jane M. 1999. "Political Mimesis." In *Collecting Visible Evidence*, edited by Jane M. Gaines and Michael Renov, 84–102. Minneapolis: University of Minnesota Press.

Gandy, Matthew. 2017. "Urban Atmospheres." *Cultural Geographies* 24, no. 3 (July): 353–74. https://doi.org/10.1177/1474474017712995.

Garbus, Liz, dir. 2021. *Becoming Cousteau*. National Geographic Documentary Films.

Garrard, Greg. 2012. "Nature Cures? Or How to Police Analogies of Personal and Ecological Health." *Interdisciplinary Studies in Literature and Environment* 19, no. 3 (Summer): 494–514.

Garrett, B. L., and A. Fish. 2016. "Attack on the Drones: The Creeping Privatisation of Our Urban Airspace." *Guardian*, December 12, 2016. https://www.theguardian.com/cities/2016/dec/12/attack-drones-privatisation-urban-airspace.

Gartry, Laura. 2018. "Robots Ready to Start Killing Crown-of-Thorns Starfish on Great Barrier Reef." ABC News, August 30, 2018. https://www.abc.net.au/news/2018-08-31/crown-of-thorns-starfish-killing-robot-great-barrier-reef-qld/10183072.

Geertz, Clifford. 1973. *The Interpretation of Cultures: Selected Essays*. New York: Basic Books.

Gelatt, Thomas S., and Roger Gentry. 2018. "Northern Fur Seal: *Callorhinus ursinus*." In *Encyclopedia of Marine Mammals*, edited by Bernd Würsig, J. G. M. Thewissen, and Kit M. Kovacs, 645–48. New York: Academic Press.

Geoghegan, J. L., V. Pirotta, E. Harvey, A. Smith, J. P. Buchmann, M. Ostrowski, J. S. Eden, et al. 2018. "Virological Sampling of Inaccessible Wildlife with Drones." *Viruses* 10 (6): 300.

Gerrard, Philippa. 2021. "How Drones Are Changing the Way We Catch Salmon Poachers." *Press and Journal*, October 5, 2021. https://www.pressandjournal.co.uk/fp/news/aberdeen-aberdeenshire/3516249/how-drones-are-changing-the-way-we-catch-salmon-poachers/.

Ghosh, Iman. 2020. "World Oceans Day: Visualizing Our Impact on Our Ocean Economy." World Economic Forum, June 8, 2020. https://www.weforum.org/agenda/2020/06/human-impact-ocean-economy.

Gibbs, Leah. 2018. "Sharks, Nets and Disputed Territory in Eastern Australia." In *Territory beyond Terra*, edited by Kimberly Peters, Philip Steinberg, and Elaine Stratford, 203–19. London: Rowman and Littlefield International.

Gibbs, Leah. 2021. "Agency in Human–Shark Encounter." *Environment and Planning E: Nature and Space* 4 (2): 645–66. https://doi.org/10.1177/2514848620929942.

Gibbs, Leah, Lachlan Fetterplace, Matthew Rees, and Quentin Hanich. 2020. "Effects and Effectiveness of Lethal Shark Hazard Management: The Shark

Meshing (Bather Protection) Program, NSW, Australia." *People and Nature* 2, no. 1 (March): 189–203.

Gibbs, Leah, and Andrew Warren. 2014. "Killing Sharks: Cultures and Politics of Encounter and the Sea." *Australian Geographer* 45, no. 2 (April): 101–7. https://doi.org/10.1080/00049182.2014.899023.

Giggs, R. 2020. *Fathoms: The World in the Whale*. Melbourne: Scribe.

Giles, A., J. E. Davies, K. Ren, and B. Kelaher. 2020. "Deep Learning Algorithms to Detect and Remove Sun Glint from High-Resolution Aerial Imagery." Abstract. American Geophysical Union, Fall Meeting, December 2020.

Gill, David A., Michael B. Mascia, Gabby N. Ahmadia, Louise Glew, Sarah E. Lester, Megan Barnes, Ian Craigie, et al. 2017. "Capacity Shortfalls Hinder the Performance of Marine Protected Areas Globally." *Nature* 543:665–69. https://doi.org/10.1038/nature21708.

Giraud, Eva. 2019. *What Comes after Entanglement*. Durham, NC: Duke University Press.

Goldfinch, Shaun, Ross Taplin, and Robin Gauld. 2021. "Trust in Government Increased during the COVID-19 Pandemic in Australia and New Zealand." *Australian Journal of Public Administration* 80, no. 1 (March): 3–11. https://doi.org/10.1111/1467-8500.12459.

Goodall, Jane. 2015. "Caring for People and Valuing Forests in Africa." In *Protecting the Wild: Parks and Wilderness, the Foundation for Conservation*, edited by George Wuerthner, Eileen Crist, and Tom Butler, 21–26. Washington, DC: Island.

Goodyear, Sheena. 2018. "Drone Caused Bears Distress in Viral Video, Researchers Say." *As It Happens*, CBC Radio, November 7, 2018. https://www.cbc.ca/radio/asithappens/as-it-happens-the-wednesday-edition-1.4895585/drone-caused-bears-distress-in-viral-video-researchers-say-1.4895589.

Graeber, David. 2009. *Direct Action: An Ethnography*. Chico, CA: AK.

Graham, Stephen. 2009. *Disrupted Cities: When Infrastructures Fail*. London: Routledge.

Gray, P. C., K. C. Bierlich, S. A. Mantell, A. S. Friedlaender, J. A. Goldbogen, and D. W. Johnston. 2019. "Drones and Convolutional Neural Networks Facilitate Automated and Accurate Cetacean Species Identification and Photogrammetry." *Methods in Ecology and Evolution* 10 (9): 1490–1500. https://doi.org/10.1111/2041-210X.13246.

Gray, Richard. 2020. "The Hunt for the Fish Pirates Who Exploit the Sea." *BBC Future*, June 5, 2020. https://www.bbc.com/future/article/20190213-the-dramatic-hunt-for-the-fish-pirates-exploiting-our-seas.

Great Barrier Reef Marine Park Authority 2018. "Port Curtis Coral Coast Regional TUMRA: Schedule 1." Australian Government. https://hdl.handle.net/11017/3922.

Green, M., C. Ganassin, and D. Reid. 2009. "Report into the NSW Shark Meshing (Bather Protection) Program." NSW Department of Primary Industries, Fisheries Conservation and Aquaculture Branch. https://www.sharksmart.nsw.

gov.au/__data/assets/pdf_file/0004/856165/Report-into-the-NSW-Shark-Meshing-Program.pdf.

Greene, Brian. 2020. *Until the End of Time: Mind, Matter, and Our Search for Meaning in an Evolving Universe*. New York: Alfred A. Knopf.

Gruen, Lori. 2015. *Entangled Empathy: An Alternative Ethic for Our Relationships with Animals*. New York: Lantern.

Hafeez, Sidrah, Man Sing Wong, Sawaid Abbas, Coco Y. T. Kwok, Janet Nichol, Kwon Ho Lee, Danling Tang, et al. 2018. "Detection and Monitoring of Marine Pollution Using Remote Sensing Technologies." In *Monitoring of Marine Pollution*, edited by Houma Bachari Fouzia, 7–32. London: IntechOpen.

Halpern, B. S., and R. R. Warner. 2002. "Marine Reserves Have Rapid and Lasting Effects." *Ecology Letters* 5 (3): 361–66. https://doi.org/10.1046/j.1461-0248.2002.00326.x.

Hamblin, Jacob D., ed. 2013. "Roundtable Review of Wired Wilderness: Technologies of Tracking and the Making of Modern Wildlife, by Etienne Benson." *h-Net Environment Roundtable Reviews* 3 (1). https://networks.h-net.org/system/files/contributed-files/env-roundtable-3-1.pdf.

Hamilton, C. 2016. "The Theodicy of the 'Good Anthropocene.'" *Environmental Humanities* 7 (1): 233–38.

Hamilton, C. 2017. *Defiant Earth: The Fate of Humans in the Anthropocene*. Cambridge: Polity.

Hansen, Jeff. 2020. "Our Clients Are Whales—That's Who We Represent." Sea Shepherd Australia website, August 12, 2020. https://www.seashepherd.org.au/latest-news/whales-clients/.

Haraway, Donna. 1988. "Situated Knowledges: The Science Question in Feminism and the Privilege of Partial Perspective." *Feminist Studies* 14 (3): 575–99.

Haraway, Donna. 1997. *Modest_Witness@Second_Millennium: FemaleMan_Meets_OncoMouse: Feminism and Technoscience*. New York: Routledge.

Haraway, Donna. 2003. *The Companion Species Manifesto: Dogs, People, and Significant Otherness*. Chicago: Prickly Paradigm.

Haraway, Donna. 2008. *When Species Meet*. Chicago: Prickly Paradigm.

Haraway, Donna. 2012. "Awash in Urine: DES and Premarin® in Multispecies Response-ability." *Women's Studies Quarterly* 40 (1/2): 301–16. http://www.jstor.org/stable/23333460.

Haraway, Donna. 2016. *Staying with the Trouble: Making Kin in the Chthulucene*. Durham, NC: Duke University Press.

Haraway, Donna. 2018. "Staying with the Trouble for Multispecies Environmental Justice." *Dialogues in Human Geography* 8, no. 1 (February): 102–5.

Hardshell Labs. N.d. "Hardshell Labs Services: Effective Asset Protection from Bird Damage—Humane Solutions." Overview. Accessed July 7, 2023. https://hardshelllabs.com/overview.

Hartshorne, J. K., Y. T. Huang, P. M. L. Paredes, K. Oppenheimer, P. T. Robbins, and M. D. Velasco. 2021. "Screen Time as an Index of Family Distress." *Current Research in Behavioral Sciences* 2. https://doi.org/10.1016/j.crbeha.2021.100023.

Harvey, David. 1996. *Justice, Nature and the Geography of Difference*. Oxford: Basil Blackwell.

Hasty, William, and Kimberley Peters. 2012. "The Ship in Geography and the Geographies of Ships." *Geography Compass* 6 (11): 660–76.

Hayward, Eva. 2010. "Fingereyes: Impressions of Cup Corals." *Cultural Anthropology* 25, no. 4 (November): 577–99.

Heidegger, Martin. (1927) 1962. *Being and Time*. Translated by John Macquarrie and Edward Robinson. London: SCM.

Heller, Peter. 2007. *The Whale Warriors: The Battle at the Bottom of the World to Save the Planet's Largest Mammals*. New York: Simon and Schuster.

Helmreich, Stefan. 2007. "An Anthropologist Underwater: Immersive Soundscapes, Submarine Cyborgs, and Transductive Ethnography." *American Ethnologist* 34:621–41.

Helmreich, Stefan. 2009a. *Alien Ocean: Anthropological Voyages in Microbial Seas*. Berkeley: University of California Press.

Helmreich, Stefan. 2009b. "Intimate Sensing." In *Simulation and Its Discontents*, edited by Sherry Turkle, 129–50. Cambridge, MA: MIT Press.

Helmreich, Stefan. 2016. *Sounding the Limits of Life: Essays in the Anthropology of Biology and Beyond*. Princeton, NJ: Princeton University Press.

Helmreich, Stefan. 2019. "Reading a Wave Buoy." *Science, Technology, and Human Values*, 44 (5): 737–61.

Henshilwood, C. S., J. Sealy, R. Yates, K. Cruz-Uribe, P. Goldberg, F. E. Grine, R. Klein, et al. 2001. "Blombos Cave, Southern Cape, South Africa: Preliminary Report on the 1992–1999 Excavations of the Middle Stone Age Levels." *Journal of Archaeological Science* 28, no. 4 (April): 421–48. https://doi.org/10.1006/JASC.2000.0638.

Herskovits, Melville J. 1955. *Cultural Anthropology*. New York: Knopf.

Heron, Scott F. 2018. "Impacts of Climate Change on World Heritage Coral Reefs: Update to the First Global Scientific Assessment." UNESDOC Digital Library. https://unesdoc.unesco.org/ark:/48223/pf0000265625.

Hildebrand, Julia M. 2019. "Consumer Drones and Communication on the Fly." *Mobile Media and Communication* 7(3): 395–411. https://doi.org/10.1177/2050157919850603.

Hildebrand, Julia M. 2021. *Aerial Play: Drone Medium, Mobility, Communication, and Culture*. Singapore: Palgrave Macmillan.

Hodder, I. 2018. *The Evolution of Humans and Things*. New Haven, CT: Yale University Press.

Hoelzl, Ingrid, and Rémi Marie. 2016. "Posthuman Vision." In *ISEA2016, Hong Kong, Cultural R>Evolution: Proceedings of the 22nd Symposium on Electronic Art*, edited by Olli Tapio Leino, 294–96. Hong Kong: University of Hong Kong.

Howard, Brian Clark. 2016. "New Galápagos Sanctuary Has World's Highest Abundance of Sharks." *National Geographic*, March 21, 2016. https://www.nationalgeographic.com/animals/article/160321-galapagos-marine-reserve-park-ecuador-conservation.

Howe, Cymene. 2019a. *Ecologics: Wind and Power in the Anthropocene*. Durham, NC: Duke University Press.

Howe, Cymene, and Dominic Boyer. 2015. "Aeolian Politics." *Distinktion: Journal of Social Theory* 16 (1): 31–48.

Hulme, Mike. 2009. *Why We Disagree about Climate Change: Understanding Controversy, Inaction and Opportunity*. Cambridge: Cambridge University Press.

Huntington, Henry P., Seth L. Danielson, Francis K. Wiese, Matthew Baker, Peter Boveng, John J. Citta, Alex De Robertis, et al. 2020. "Evidence Suggests Potential Transformation of the Pacific Arctic Ecosystem Is Underway." *Nature Climate Change* 10 (April): 342–48.

Ianelli, James N., S. J. Barbeaux, and D. McKelvey. 2020. "Assessment of Walleye Pollock in the Bogoslof Island Region." Alaska Fisheries Science Center and National Marine Fisheries Service, December 2020. https://apps-afsc .fisheries.noaa.gov/refm/docs/2020/BOGpollock.pdf.

Ingalsbee, Timothy. 1996. "Earth First! Activism: Ecological Postmodern Praxis in Radical Environmentalist Identities." *Sociological Perspectives* 39 (2): 263–76.

Ingersoll, Karin Amimoto.. 2016. *Waves of Knowing: A Seascape Epistemology*. Durham, NC: Duke University Press.

Ingold, T. 2007. "Earth, Sky, Wind, and Weather." *Journal of the Royal Anthropological Institute* 13, no. S1 (April): S19–S38. https://doi.org/10.1111 /j.1467-9655.2007.00401.x.

Ingold, Timothy. 2008. "Bindings against Boundaries: Entanglements of Life in an Open World." *Environment and Planning A: Economy and Space* 40, no. 8 (August): 1796–1810.

Ingold, Timothy. 2010. "Footprints through the Weather-World: Walking, Breathing, Knowing." *Journal of the Royal Anthropological Institute* 16, no. S1 (May): S121–39.

Inoue, J., and J. A. Curry. 2004. "Application of Aerosondes to High-Resolution Observations of Sea Surface Temperature over Barrow Canyon." *Geophysical Research Letters* 31, no. 14 (July). https://doi.org/10.1029/2004GL020336.

IUCN. 2013. "World Nearing 3% of Ocean Protection." International Union for Conservation of Nature, October 23, 2013. https://www.iucn.org/content /world-nearing-3-ocean-protection.

Jablonowski, Maximilian. 2020. "Beyond Drone Vision: The Embodied Telepresence of First-Person-View Drone Flight." *Senses and Society* 15, no. 3 (October): 344–58. https://doi.org/10.1080/17458927.2020.1814571.

Jackman, Anna. 2016. "Rhetorics of Possibility and Inevitability in Commercial Drone Tradescapes." *Geographica Helvetica* 71 (1): 1–6. https://doi.org/10.5194 /gh-71-1-2016.

Jackson, Steven. 2014. "Rethinking Repair." In *Media Technologies: Essays on Communication, Materiality, and Society*, edited by Tarleton Gillespie, Pablo Boczkowski, and Kirsten Foot, 221–40. Cambridge, MA: MIT Press.

Jiménez, Alberto Corsín, and Chloe Nahum-Claudel. 2019. "The Anthropology of

Traps: Concrete Technologies and Theoretical Interfaces." *Journal of Material Culture* 24, no. 4 (January): 383–400.

Johnston, D. W. 2019. "Unoccupied Aircraft Systems in Marine Science and Conservation." *Annual Review of Marine Science* 11 (1): 439–63.

Jones, Kendall R., Carissa J. Klein, Hedley S. Grantham, Hugh P. Possingham, Benjamin S. Halpern, Neil D. Burgess, Stuart H. M. Butchart, et al. 2020. "Area Requirements to Safeguard Earth's Marine Species." *One Earth* 2, no. 2 (February): 188–96. https://doi.org/10.1016/j.oneear.2020.01.010.

Joyce, Karen E., Stephanie Duce, Susannah Leahy, Javier Leon, and Stefan Maier. 2018. "Principles and Practice of Acquiring Drone-Based Image Data in Marine Environments." *Marine and Freshwater Research* 70 (7): 952–63. https://doi.org/10.1071/MF17380.

Jue, Melody. 2020. *Wild Blue Media: Thinking through Seawater.* Durham, NC: Duke University Press.

Jue, Melody, and Rafico Ruiz, eds. 2021. *Saturation: An Elemental Politics.* Durham, NC: Duke University Press.

Kaplan, Caren. 2017. "Drone-o-Rama: Troubling the Spatial and Temporal Logics of Distance Warfare." In *Life in the Age of Drone Warfare*, edited by Lisa Parks and Caren Kaplan, 161–77. Durham, NC: Duke University Press.

Kapp, Ernst. (1877) 2018. *Elements of a Philosophy of Technology: On the Evolutionary History of Culture.* Edited by Jeffrey West Kirkwood and Leif Weatherby, translated by Lauren K. Wolfe. Minneapolis: University of Minnesota Press.

Kays, Roland, William J. McShea, and Martin Wikelski. 2020. "Born-Digital Biodiversity Data: Millions and Billions." *Diversity and Distributions* 26 (5): 644–48.

Kelaher, Brendan P., Andrew P. Colefax, Alejandro Tagliafico, Melanie J. Bishop, Anna Giles, and Paul Butcher. 2020. "Assessing Variation in Assemblages of Large Marine Fauna Off Ocean Beaches Using Drones." *Marine and Freshwater Research* 71 (1): 68–77.

Keller, Bryn. 2018. "Machine Learning, Drones, and Whales: A Great Combination!" *Xoltar: An Online Notebook by Bryn Keller.* April 26, 2018. http://www.xoltar.org/posts/2018-04-26-whale-expedition/index.html.

Kelly, Mark. 2009. *The Political Philosophy of Michel Foucault.* New York: Routledge.

Kenner, Alison, Aftab Mirzaei, and Christy Spackman. 2019. "Breathing in the Anthropocene: Thinking through Scale with Containment Technologies." *Cultural Studies Review* 25, no. 2 (December): 153–71. https://doi.org/10.5130/csr.v25i2.6941.

Kerr, Iain. N.d.a. "Parley SnotBot, Alaska Expedition Powered by Intel: Stunned in se Alaska." Ocean Alliance. Accessed August 18, 2020. https://whale.org/parley-snotbot-alaska-expedition-powered-intel-stunned-se-alaska/.

Kerr, Iain. N.d.b. "SnotBot Sea of Cortez Part 3: Blue Whale." Ocean Alliance. Accessed August 18, 2020. https://whale.org/snotbot-sea-of-cortez-part-3-blue-whale/.

Khatchadourian, Raffi. 2007. "Neptune's Navy." *New Yorker*, October 29, 2007. https://www.newyorker.com/magazine/2007/11/05/neptunes-navy.

Kieza, Grantlee. 2020. *Banks*. Sydney: Harper Collins.

Kinzelbach, Ragnar. 1986. "The Sperm Whale, *Physeter macrocephalus*, in the Eastern Mediterranean Sea." *Zoology in the Middle East* 1 (1): 15–17. https://doi.org /10.1080/09397140.1986.11770900.

Kirksey, Eben. 2015. *Emergent Ecologies*. Durham, NC: Duke University Press.

Kirksey, Eben. 2019. "Queer Love, Gender Bending Bacteria, and Life after the Anthropocene." *Theory, Culture and Society* 36 (6): 197–219.

Kittler, Fredrich. 2009. "Toward an Ontology of Media." *Theory, Culture, and Society* 26 (2–3): 23–31.

Kituyi, Mukhisa, and Peter Thomson. 2018. "90% of Fish Stocks Are Used Up— Fisheries Subsidies Must Stop." World Economic Forum website, July 13, 2018. https://www.weforum.org/agenda/2018/07/fish-stocks-are-used-up- fisheries-subsidies-must-stop/.

Kleinberg-Levin, David M. 1984. "Logos and Psyche: A Hermeneutics of Breathing." *Research in Phenomenology* 14 (1): 121–47.

Koh, Lian Pin. N.d. "Poachers Expected to Use Green Drones to Kill Endangered Wildlife." Ecological Society of Australia. Accessed May 5, 2023. https://www .ecolsoc.org.au/news/media/poachers-expected-to-use-green-drones-to -kill-endangered-wildlife/.

Kohn, Eduardo. 2013. *How Forests Think: Toward an Anthropology beyond the Human*. Berkeley: University of California Press.

Kopnina, Helen. 2016. "Nobody Likes Dichotomies (but Sometimes You Need Them)." *Anthropological Forum* 26 (4): 415–29.

Kopnina,Helen. 2017. "Beyond Multispecies Ethnography: Engaging with Violence and Animal Rights in Anthropology." *Critique of Anthropology* 37 (3): 333–57.

Kottak, Conrad P. 1999. "The New Ecological Anthropology." *American Anthropologist* 101 (1): 23–35.

Kuhn, C. 2017. "Northern Fur Seal Food Study—Post 6." NOAA *Science Blog*, November 20, 2017. https://www.fisheries.noaa.gov/science-blog/northern -fur-seal-food-study-post-6.

Kuhn, C., A. De Robertis, J. Sterling, and C. W. Mordy. 2020. "Test of Unmanned Surface Vehicles to Conduct Remote Focal Follow Studies of a Marine Predator." *Marine Ecology Progress Series*, no. 635, 1–7.

Ladkani, R., dir. 2019. *Sea of Shadows*. Vienna: Terra Mater Factual Studios. https://www.austrianfilms.com/film/sea_of_shadows.

Lang, M. 2020. "Leaf-Blower Wars: How Portland Protesters Are Fighting Back against Tear Gas and Forming 'Walls' of Veterans, Lawyers, Nurses." *Washington Post*, July 26, 2020. https://www.washingtonpost.com/nation/2020/07/26 /leaf-blower-wars-how-portland-protesters-are-fighting-back-against-tear -gas-forming-walls-veterans-lawyers-nurses/.

Latimer, Joanna. 2013. "Being Alongside: Rethinking Relations amongst Different Kinds." *Theory, Culture, and Society* 30 (7–8): 77–104.

Latimer, Joanna, and Mara Miele. 2013. "Naturecultures? Science, Affect and the Non-human." *Theory, Culture and Society* 30 (7–8): 5–31.

Latour, Bruno. 1993. *We Have Never Been Modern*. Cambridge, MA: Harvard University Press.

Latour, Bruno. 2004. *The Politics of Nature: How to Bring the Sciences into Democracy*. Cambridge, MA: Harvard University Press.

Latour, Bruno. 2017. *Facing Gaia: Eight Lectures on the New Climatic Regime*. Cambridge: Polity.

Latour, Bruno, and Peter Weibel., eds. 2005. *Making Things Public: Atmospheres of Democracy*. Cambridge, MA: MIT Press.

Latour, Bruno, and Steve Woolgar. 1986. *Laboratory Life: The Construction of Scientific Facts*. 2nd ed. Princeton, NJ: Princeton University Press.

Law, John. 2004. *After Method: Mess in Social Science Research*. London: Routledge.

Lehman, Jessica. 2013. "Relating to the Sea: Enlivening the Ocean as an Actor in Eastern Sri Lanka." *Environment and Planning D: Society and Space* 31 (3): 485–501. https://doi.org/10.1068/d24010.

Lehman, Jessica. 2016. "A Sea of Potential: The Politics of Global Ocean Observations." *Political Geography* 55 (November): 113–23. https://doi.org/10.1016/j.polgeo.2016.09.006.

Lehman, Jessica. 2017. "From Ships to Robots: The Social Relations of Sensing the World Ocean." *Social Studies of Science* 48 (1): 57–79.

Lehman, Jessica, Philip Steinberg, and Elizabeth R. Johnson. 2021. "Turbulent Waters in Three Parts." *Theory and Event* 24 (1): 192–219.

Lemke, Thomas. 2015. "New Materialisms: Foucault and the 'Government of Things.'" *Theory, Culture and Society* 32 (4): 3–25.

Leroi-Gourhan, André. (1964) 1993. *Gesture and Speech*. Translated by A. B. Berger. Cambridge, MA: MIT Press.

Liboiron, Max. 2021. *Pollution Is Colonialism*. Durham, NC: Duke University Press.

Linchant, Julie, Jonathan Lisein, Jean Semeki Ngabinzeke, Philippe Lejeune, and Cédric Vermeulen. 2015. "Are Unmanned Aircraft Systems (UASs the Future of Wildlife Monitoring? A Review of Accomplishments and Challenges." *Mammal Review* 45 (4): 239–52. https://doi.org/10.1111/mam.12046.

Lorimer, Jamie. 2015. *Wildlife in the Anthropocene*. Minneapolis: University of Minnesota Press.

Lorimer, Jamie. 2017. "Parasites, Ghosts and Mutualists: A Relational Geography of Microbes for Global Health." *Transactions of the Institute of British Geographers* 42 (4): 544–58.

Lovelock, James. 2019. *Novacene: The Coming Age of Hyperintelligence*. London: Allen Lane.

Lovink, Geert. 2005. *The Principle of Notworking.* Amsterdam: Amsterdam University Press.

Lovink, Geert, and Joanna Richardson. 2001. "Notes on Sovereign Media." Subsol, November 2001. http://subsol.c3.hu/subsol_2/contributors0/lovink-richardsontext.html.

Lowe, Celia. 2010. "Viral Clouds: Becoming H5N1 in Indonesia." *Cultural Anthropology* 25, no. 4 (November): 625–49. https://doi.org/10.1111/j.1548-1360.2010.01072.x.

Lowery, Wesley. 2016. *They Can't Kill Us All: Ferguson, Baltimore, and a New Era in America's Racial Justice Movement.* New York: Little, Brown.

Lowry, Lloyd F., Kathryn J. Frost, and Thomas R. Loughlin. 1988. "Importance of Walleye Pollock in the Diets of Marine Mammals in the Gulf of Alaska and Bering Sea, and Implications for Fishery Management." In *Proceedings of the International Symposium on the Biology and Management of Walleye Pollock,* November 14–16, pp. 701–26. Anchorage: Alaska Sea Grant. https://www.adfg.alaska.gov/static/home/library/pdfs/wildlife/research_pdfs/88_lowry_etal_walleye_pollock_diets_of_marine_mammals.pdf.

Luhmann, Niklas. 1986. "The Autopoiesis of Social Systems." *Sociocybernetic Paradoxes* 6 (2): 172–92.

Luke. 2014. "Catch and Release WA Tiger Shark Cull." YouTube video, March 27, 2014. https://www.youtube.com/watch?v=yR_PBnIgoMw.

Lunstrum, Elizabeth. 2014. "Green Militarization: Anti-poaching Efforts and the Spatial Contours of Kruger National Park." *Annals of the Association of American Geographers* 104 (4): 816–32.

Macauley, David. 2010. *Elemental Philosophy: Earth, Air, Fire, and Water as Environmental Ideas.* Albany: State University of New York Press.

Malm, Andreas. 2018. *The Progress of This Storm: Nature and Society in a Warming World.* London: Verso.

Marcus Aurelius Value. 2017. "Pingtan Marine: A Fraud That Finances Human Trafficking and Poaching." May 10, 2017. http://www.mavalue.org/research/pingtan-marine-fraud-finances-human-trafficking-poaching/.

Marean, Curtis W. 2012. "When the Sea Saved Humanity." *SA Special Editions* 22, no. 1 (December): 52–59. https://doi.org/10.1038/scientificamericanhuman1112-52.

Marine Conservation Institute. N.d. *Marine Protection Atlas.* Accessed March 2, 2022. https://mpatlas.org/.

Marshall, Greg J. 1998. "Crittercam: An Animal-Borne Imaging and Data Logging System." *Marine Technology Society Journal* 32:11–17.

Marshall, Greg. 2011. "PSW 2295 Crittercam Animal-Borne Imaging." PSW Science, YouTube video, January 11, 2011. https://www.youtube.com/watch?v=JqcWZfVJHRM.

Marx, Karl. (1867) 1992. *Capital.* New York: Penguin.

Massé, Francis. 2018. "Topographies of Security and the Multiple Spatialities of

(Conservation) Power: Verticality, Surveillance, and Space-Time Compression in the Bush." *Political Geography* 67 (November), 56–64.

Massey, Doreen. 2004. "Geographies of Responsibility." *Geografiska Annaler: Series B, Human Geography* 86 (1): 5–18.

Matheson, Rob. 2019. "A Battery-Free Sensor for Underwater Exploration." *MIT News*, August 20. https://news.mit.edu/2019/battery-free-sensor-underwater -exploration-0820.

Maxwell, Sara, Elliott Hazen, Rebecca Lewison, Daniel Dunn, Helen Bailey, Steven Bograd, Dana Briscoe, et al. 2015. "Dynamic Ocean Management: Defining and Conceptualizing Real-Time Management of the Ocean." *Marine Policy* 58:42–50. https://doi.org/10.1016/j.marpol.2015.03.014.

McCormack, Derek P. 2016. "Elemental Infrastructures for Atmospheric Media: On Stratospheric Variations, Value and the Commons." *Environment and Planning D: Society and Space* 35 (3): 318–437.

McCormack, Derek P. 2018. *Atmospheric Things: On the Allure of Elemental Envelopment*. Durham, NC: Duke University Press.

McEvoy, John F., Graham P. Hall, and Paul G. McDonald. 2016. "Evaluation of Unmanned Aerial Vehicle Shape, Flight Path and Camera Type for Waterfowl Surveys: Disturbance Effects and Species Recognition." *Peer Journal* 4:e1831. https://doi.org/10.7717/peerj.1831.

McIntyre, E. M., and A. J. Gasiewski. 2007. "An Ultra-lightweight L-Band Digital Lobe-Differencing Correlation Radiometer (LDCR) for Airborne UAV SSS Mapping." In *2007 IEEE International Geoscience and Remote Sensing Symposium*, 1095–97. New York: IEEE.

McLaren, Duncan, Daryl Fedje, Angela Dyck, Quentin Mackie, Alisha Gauvreau, and Jenny Cohen. 2018. "Terminal Pleistocene Epoch Human Footprints from the Pacific Coast of Canada." *PLOS ONE* 13, no. 3 (March): e0193522. https://doi .org/10.1371/journal.pone.0193522.

McLuhan, Marshall. 1964. *Understanding Media: The Extensions of Man*. New York: McGraw-Hill.

McLuhan, Marshall. 1969. "The Playboy Interview: Marshall McLuhan." *Playboy*, March 1969.

McNeill, Donald. 2020. "The Volumetric City." *Progress in Human Geography* 44, no. 5 (October): 815–31. https://doi.org/10.1177/0309132519863486.

McPhee, Daryl. 2014. "Unprovoked Shark Bites: Are They Becoming More Prevalent?" *Coastal Management* 42 (5): 478–92. https://doi.org/10.1080/08920753.2 014.942046.

McPhee, Daryl, Craig Blount, Will MacBeth, and Dilys Zhang. 2019. "Queensland Shark Control Program, Cardno." Queensland Government. https://www .publications.qld.gov.au/dataset/e20e6bcd-c076-42a2-9e17-7d549b02254e /resource/76358bc5-a2fa-46ce-a8cb-0891c75e971a/download/qld-shark -control-program-review-alternative-approaches.pdf.

Miller, James. 1993. *The Passion of Michel Foucault*. London: Harper Collins.

Millner, Naomi. 2020. "As the Drone Flies: Configuring a Vertical Politics of Contestation within Forest Conservation." *Political Geography* 80:102–63.

Mirzoeff, Nicholas. 2009. "The Sea and the Land: Biopower and Visuality from Slavery to Katrina." *Culture, Theory and Critique* 50, no. 2–3 (December): 289–305. https://doi.org/10.1080/14735780903240331.

Moffa, Anthony. 2012. "Two Competing Models of Activism, One Goal: A Case Study of Anti-whaling Campaigns in the Southern Ocean." *Yale Journal of International Law* 37 (1): 201–14.

Mordy, C. W., E. D. Cokelet, A. De Robertis, R. Jenkins, C. E. Kuhn, N. Lawrence-Slavas, C. L. Berchok, et al. 2017. "Advances in Ecosystem Research: Saildrone Surveys of Oceanography, Fish, and Marine Mammals in the Bering Sea." *Oceanography* 30 (2): 113–15. http://www.jstor.org/stable/26201857.

Mulero-Pázmány, Margarita, Roel Stolper, L. D. van Essen, Juan J. Negro, and Tyrell Sassen. 2014. "Remotely Piloted Aircraft Systems as a Rhinoceros Anti-poaching Tool in Africa." *PLOS ONE* 9 (1): e83873. https://doi.org/10.1371/journal.pone.0083873.

Murphy, Michelle. 2008. "Chemical Regimes of Living." *Environmental History* 13 (4): 695–703.

Naess, Arne. 1973. "The Shallow and the Deep, Long-Range Ecology Movement: A Summary." *Inquiry* 16 (1–4): 95–100. https://doi.org/10.1080/00201747308601682.

Nagtzaam, Gerry. 2014. "Gaia's Navy: The Sea Shepherd Conservation Society's Battle to Stay Afloat and International Law." *William and Mary Environmental Law and Policy Review* 38 (3): 613–94.

Nagtzaam, Gerry, and Douglas Guilfoyle. 2018. "'Ramming Speed': The Sea Shepherd Conservation Society and the Law of Protest." *Monash University Law Review* 44 (2): 360–83.

Nail, T. 2021. *Theory of the Earth.* Stanford, CA: Stanford University Press.

Nealon, Sean. 2019. "Using Drones, GoPros to Track Gray Whale Behavior, and Spot Their Poop, Off Oregon Coast." Oregon State University Newsroom, November 18, 2019. https://today.oregonstate.edu/news/using-drones-gopros-track-gray-whale-behavior-and-spot-their-poop-oregon-coast.

Neimanis, Astrid. 2017. *Bodies of Water: Posthuman Feminist Phenomenology.* New York: Bloomsbury.

Neimark, Benjamin. 2019. "Address the Roots of Environmental Crime." *Science* 364, no. 6436 (April): 139.

Nicol, Christine, Lars Bejder, Laura Green, Craig Johnson, Linda Keeling, Dawn Noren Dawn, Julie Van der Hoop, et al. 2020. "Anthropogenic Threats to Wild Cetacean Welfare and a Tool to Inform Policy in This Area." *Frontiers in Veterinary Science* 7:57.

NOAA. N.d. "How Important Is the Ocean to Our Economy?" National Oceanic and Atmospheric Administration. Accessed March 2, 2022. https://oceanservice.noaa.gov/facts/oceaneconomy.html.

NOAA. 2019. "Fisheries of the Exclusive Economic Zone off Alaska; Bering Sea and Aleutian Islands; Final 2019 and 2020 Harvest Specifications for Groundfish." *Federal Register* 84, 49 (March 13, 2019). https://www.federalregister.gov /documents/2019/03/13/2019-04539/fisheries-of-the-exclusive-economic -zone-off-alaska-bering-sea-and-aleutian-islands-final-2019.

Noble, Safiya Umoja. 2018. *Algorithms of Oppression: How Search Engines Reinforce Racism*. New York: New York University Press.

O'Connor, Sue, Ono Rintaro, and Chris Clarkson. 2011. "Pelagic Fishing at 42,000 Years before the Present and the Maritime Skills of Modern Humans." *Science* 334, no. 6059 (November): 1117–21. https://doi.org/10.1126/science .1207703.

Ogborn, Miles. 2000. "Historical Geographies of Globalisation." In *Modern Historical Geographies*, edited by Catherine Nash and Brian. J. Graham, 43–69. Harlow, UK: Pearson/Longman.

O'Grady, Nathaniel. 2019. "Communication and the Elemental: Capacities, Force and Excess in Emergency Information Sharing." *Environment and Planning D: Society and Space* 37 (1): 158–76.

Ong, Thuy. 2017. "Dutch Police Will Stop Using Drone-Hunting Eagles Since They Weren't Doing What They're Told." *Verge*, December 12, 2017. https://www .theverge.com/2017/12/12/16767000/police-netherlands-eagles-rogue-drones.

Oppermann, Serpil. 2019. "Storied Seas and Living Metaphors in the Blue Humanities." *Configurations* 27 (4): 443–61.

Oreskes, N. 2021. *Science on a Mission*. Chicago: University of Chicago Press.

Pacific Marine Environmental Laboratory. 2018. "Follow the Saildrone 2018: Monitoring a Changing Arctic." October 8, 2018. https://www.pmel.noaa.gov /itae/follow-saildrone-2018.

Palmer, Clare. 2016. "Taming the Wild Profusion of Existing Things? A Study of Foucault, Power and Animals." In *Foucault and Animals*, edited by Matthew Chrulew and Dinesh Wadiwal, 105–31. Leiden: Brill Academic.

Pancia, A. 2019. "Blue Whale, World's Largest Animal, Caught on Camera Having a Poo." ABC News, November 15, 2019. https://www.abc.net.au/news/2019 -11-16/blue-whale-worlds-largest-animal-caught-on-camera-having-a-poo /11708368.

Papadopoulos, D., M. Puig de la Bellacasa, and N. Myers. 2022. *Reactivating Elements: Chemistry, Ecology, Practice*. Durham, NC: Duke University Press.

Papadopoulos, J., and D. Ruscillo. 2002. "A Ketos in Early Athens: An Archaeology of Whales and Sea Monsters in the Greek World." *American Journal of Archaeology* 106 (2): 187–227. https://doi.org/10.2307/4126243.

Parikka, Jussi. 2015. *A Geology of Media*. Minneapolis: University of Minnesota Press.

Parks, Lisa. 2009. "Around the Antenna Tree: The Politics of Infrastructural Visibility." *Flow*, March 6, 2009. http://www.flowjournal.org/2009/03/around -the-antenna-tree-the-politics-of-infrastructural-visibilitylisa-parks-uc -santa-barbara/.

Parks, Lisa. 2017. "Vertical Mediation and the U.S. Drone War in the Horn of Africa." In *Life in the Age of Drone Warfare*, edited by Lisa Parks and Caren Kaplan, 134–58. Durham, NC: Duke University Press.

Paterson, Mark. 2006. "Feel the Presence: Technologies of Touch and Distance." *Environment and Planning D: Society and Space* 24 (5): 691–708.

Paxson, H. 2008. "Post-Pasteurian Cultures: The Microbiopolitics of Raw-Milk Cheese in the United States." *Cultural Anthropology* 23:15–47.

Paxson, H., and S. Helmreich. 2014. "The Perils and Promises of Microbial Abundance: Novel Natures and Model Ecosystems, from Artisanal Cheese to Alien Seas." *Social Studies of Science* 44 (2): 165–93. http://doi.org/10.1177 /0306312713505003.

Payne, Roger, producer. 1970. *Songs of the Humpback Whale*. CRM Records SWR 11, LP, 34:26.

Payne, Roger. N.d. "Aerial and Underwater Drones." Ocean Alliance. Accessed May 5, 2023. https://whale.org/aerial-underwater-drones-roger-payne/.

Peckham, Robert, and Ria Sinha. 2019. "Anarchitectures of Health: Futures for the Biomedical Drone." *Global Public Health* 14 (8): 1204–19. https://doi.org /10.1080/17441692.2018.1546335.

Pedelty, Mark. 1995. *War Stories: The Culture of Foreign Correspondents*. New York: Routledge.

Pendergast, Joanna. 2017. "Wedge-Tailed Eagle Takes Down Drone Flying over West Australian Wheat Farm." ABC News, May 24, 2017. https://www.abc.net .au/news/rural/2017-05-24/wedge-tailed-eagle-takes-down-drone-over- wa-wheat-farm/8554120.

Pepin-Neff, Christopher. 2012. "Australian Beach Safety and the Politics of Shark Attacks." *Coast Management* 40, no. 1 (January): 88–106.

Perrow, Charles. 1984. *Normal Accidents: Living with High-Risk Technologies*. New York: Basic Books.

Pershing, Andrew J., Line B. Christensen, Nicholas R. Record, Graham D. Sherwood, and Peter. B. Stetson. 2010. "The Impact of Whaling on the Ocean Carbon Cycle: Why Bigger Was Better." *PLOS ONE* 5 (8): e12444. https://doi .org/10.1371/journal.pone.0012444.

Peters, John Durham. 2015. *The Marvelous Clouds: Toward a Philosophy of Elemental Media*. Chicago: University of Chicago Press.

Peters, John Durham. 2017. "The Media of Breathing." In *Atmospheres of Breathing: Respiratory Questions of Philosophy*, edited by Lenart Škof and Petri Berndtson, 179–95. Albany: State University of New York Press.

Peters, Kimberley, and Philip Steinberg. 2019. "The Ocean in Excess: Towards a More-Than-Wet Ontology." *Dialogues in Human Geography* 9 (3): 293–307.

Peterson, Marina. 2017. "Atmospheric Sensibilities: Noise, Annoyance, and Indefinite Urbanism." *Social Text* 35 (2): 69–90.

Phillips, Catherine. 2017. "Discerning Ocean Plastics: Activist, Scientific, and Artistic Practices." *Environment and Planning A: Economy and Space* 49, no. 5 (January): 1146–62. https://doi.org/10.1177/0308518X16687301.

Pierschel, Marc. 2017. "184 Documentary." Hard to Port, April 14, 2017. YouTube video. https://www.youtube.com/watch?v=xw5XyFfxHqo.

Pirotta, V., A. Smith, M. Ostrowski, D. Russell, I. D. Jonsen, A. Grech, and R. Harcourt. 2017. "An Economical Custom-Built Drone for Assessing Whale Health." *Frontiers for Marine Science* 4:425.

Pitman, R. L., and J. W. Durban. 2012. "Cooperative Hunting Behavior, Prey Selectivity and Prey Handling by Pack Ice Killer Whales (*Orcinus orca*), Type B, in Antarctic Peninsula Waters." *Marine Mammal Science* 28, no. 1 (January 2012): 16–36. https://doi.org/10.1111/j.1748-692.2010.00453.x.

Plumwood, Val. 2001. *Environmental Culture: The Ecological Crisis of Reason.* London: Routledge.

Pooley, Simon, Maan Barua, William Beinart, Amy Dickman, George Holmes, Jamie Lorimer, Andrew Loveridge, et al. 2017. "An Interdisciplinary Review of Current and Future Approaches to Improving Human–Predator Relations." *Conservation Biology* 31 (June): 513–23.

Povinelli, Elizabeth. 2016. *Geontologies: A Requiem to Late Liberalism.* Durham, NC: Duke University Press.

Preston, Alana. 2012. "Eco-terrorism in the Southern Ocean: A Dangerous Byproduct of the Tangled Web of International Whaling Conventions and Treaties." *Whittier Law Review* 34:117.

Price, Leigh. 2019. "The Possibility of Deep Naturalism: A Philosophy for Ecology." *Journal of Critical Realism* 18 (4): 352–67.

Probyn, Elspeth. 2016. *Eating the Ocean.* Durham, NC: Duke University Press.

Puig de la Bellacasa, María. 2009. "Touching Technologies, Touching Visions: The Reclaiming of Sensorial Experience and the Politics of Speculative Thinking." *Subjectivity* 28:297–315.

Puig de la Bellacasa, María. 2010. "Ethical Doings in Naturecultures." *Ethics, Place and Environment* 13 (2): 151–69. https://doi.org/10.1080/13668791003778834.

Purdy, Jedediah. 2015. *After Nature: A Politics for the Anthropocene.* Cambridge, MA: Harvard University Press.

Queensland Department of Agriculture and Fisheries. 2018. "Queensland Shark Control Program Scientific Working Group Minutes." November 15, 2018. https://www.daf.qld.gov.au/business-priorities/fisheries/shark-control-program/plan/minutes-15-november-2018.

Queensland Department of Agriculture and Fisheries. 2019. "Queensland Shark Control Program, Scientific Working Group Minutes." June 14, 2019. https://www.daf.qld.gov.au/business-priorities/fisheries/shark-control-program/plan/minutes-14-june-2019.

Queensland Department of Agriculture and Fisheries. 2020. "Queensland Shark Control Program." Communique 11 and 18 September 2020. https://www.daf.qld.gov.au/business-priorities/fisheries/shark-control-program/plan/communique-11-18-september-2020.

Queensland Department of Agriculture and Fisheries. 2021a. "How We Catch and Detect Sharks: Shark Nets and Drumlines." https://www.daf.qld.gov.au

/business-priorities/fisheries/shark-control-program/shark-control
-equipment/nets-drumlines.

Queensland Department of Agriculture and Fisheries. 2021b. "Queensland
Shark Control Program." Communique 18 and 19 February 2021. https://www
.daf.qld.gov.au/business-priorities/fisheries/shark-control-program/plan
/communique-12-march-2021.

Queensland Department of Agriculture and Fisheries. 2021c. "SharkSmart Drone
Trial: Community Sentiment Report." https://www.publications.qld.gov.au
/dataset/e20e6bcd-c076-42a2-9e17-7d549b02254e/resource/f1430d5f-b111
-41dd-9403-e5a42ec81ae6/download/d_sharksmart-drone-trial-community
-sentiment-report-april-2021.pdf.

Radjawali, Irenda, Oliver Pye, and Michael Flitner. 2017. "Recognition through
Reconnaissance? Using Drones for Counter-mapping in Indonesia." *Journal of
Peasant Studies* 44, 753–69.

Ransby, Barbara. 2018. *Making All Black Lives Matter: Reimagining Freedom in the
Twenty-First Century*. Oakland: University of California Press.

Raverty, S. A., L. D. Rhodes, E. Zabek, A. Eshghi, C. Cameron, M. B. Hanson, and
J. P. Schroeder. 2017. "Respiratory Microbiome of Endangered Southern Resident Killer Whales and Microbiota of Surrounding Sea Surface Microlayer
in the Eastern North Pacific." *Science Report* 7:394. https://doi.org/10.1038
/s41598-017-00457-5.

Rawls, John. 1971. *A Theory of Justice*. Cambridge, MA: Harvard University Press.

Rebolo-Ifrán, Natalia, Maricel Graña Grilli, and Sergio Lambertucci. 2019.
"Drones as a Threat to Wildlife: YouTube Complements Science in Providing
Evidence about Their Effect." *Environmental Conservation* 46, no. 3 (June): 205–
10. http://doi.org/10.1017/S0376892919000080.

Redrobe, Karen. 2010. *Crash: Cinema and Politics of Speed and Stasis*. Durham, NC:
Duke University Press.

Reestorff, Camilla M. 2014. "Mediatised Affective Activism: The Activist Imag-
inary and the Topless Body in the Femen Movement." *Convergence* 20, no. 4
(November): 478–95.

Rekittke, Joerg, and Yazid Ninsalam. 2016. "Sliced Ecosystem: Modelling Transects of Vulnerable Marine Landscapes." *Journal of Digital Landscape Architec-
ture* 1 (July): 36–45. https://doi.org/10.14627/537612005.

Ricoeur, Paul. 1976. *Interpretation Theory: Discourse and the Surplus of Meaning*.
Fort Worth: Texas Christian University Press.

Robbins, Jim. 2018. "Orcas of the Pacific Northwest Are Starving and Disappearing." *New York Times*, July 6, 2018. https://www.nytimes.com/2018/07/09
/science/orcas-whales-endangered.html.

Robé, Chris. 2015. "The Convergence of Eco-activism, Neoliberalism, and Reality
TV in *Whale Wars*." *Journal of Film and Video* 67 (3–4): 94–111.

Roberts, Ben. 2012. "Technics, Individuation and Tertiary Memory: Bernard
Stiegler's Challenge to Media Theory." *New Formations*, no. 77 (Spring): 8–20.

Roeschke, Joseph. 2009. "Eco-terrorism and Piracy on the High Seas: Japanese

Whaling and the Rights of Private Groups to Enforce International Conservation Law in Neutral Waters." *Villanova Environmental Law Journal* 20 (1): 99–137.

Roff, George, Christopher Brown, Mark Priest, and Peter Mumby. 2018. "Decline of Coastal Apex Shark Populations over the Past Half Century." *Communications Biology* 1, no. 223 (December). https://doi.org/10.1038/s42003-018-0233-1.

Rogan, Andy. N.d. "The Sounds of a SnotBot Expedition." Ocean Alliance. Accessed August 18, 2020. https://whale.org/the-sounds-of-a-snotbot-expedition/.

Rohrlich, J. 2019. "'Tourists' Caught in US with Endangered Fish Bladders Worth $3.7 Million." *Quartz*, June 18, 2019. https://qz.com/1646387/smugglers-caught-with-3-7-million-of-endangered-totoaba-bladders/.

Roman, Joe, and James J. McCarthy. 2010. "The Whale Pump: Marine Mammals Enhance Primary Productivity in a Coastal Basin." *PLOS ONE* 5, no. 10 (October): e13255. https://doi.org/10.1371/journal.pone.0013255.

Rose, Deborah Bird. 2011. *Wild Dog Dreaming: Love and Extinction.* Charlottesville: University of Virginia Press.

Rose, Deborah Bird. 2013. "Val Plumwood's Philosophical Animism: Attentive Interactions in the Sentient World." *Environmental Humanities* 3 (1): 93–109.

Rosiek, Jerry L., Jimmy Snyder, and Scott L. Pratt. 2020. "The New Materialisms and Indigenous Theories of Non-human Agency: Making the Case for Respectful Anti-colonial Engagement." *Qualitative Inquiry* 26 (3–4): 331–46.

Rovelli, Carlo. 2018. *The Order of Time.* New York: Riverhead.

Roy, Eleanor. 2020. "'Intelligent Drones': Albatross Fitted with Radar Detectors to Spot Illegal Fishing." *Guardian*, January 30, 2020. https://www.theguardian.com/world/2020/jan/31/intelligent-drones-albatross-fitted-with-radar-detectors-to-spot-illegal-fishing.

Rutherford, Stephanie. 2007. "Green Governmentality: Insights and Opportunities in the Study of Nature's Rule." *Progress in Human Geography* 31 (3): 291–307.

Saikia, Arupjyoti. 2009. "The Kaziranga National Park: Dynamics of Social and Political History." *Conservation and Society* 7 (2): 113–29.

Saildrone. 2019. "Using Ocean Drones to Track Fur Seals in the Bering Sea: How Innovative Autonomous Technology Is Helping Scientists to Solve the Mystery of Northern Fur Seals Disappearing from the Remote Pribilof Islands." Saildrone website. March 13, 2019. https://www.saildrone.com/news/northern-fur-seals-tracking-bering-sea-2016-mission.

Saildrone. N.d. "What Is a Saildrone?" Accessed May 5, 2023. https://www.saildrone.com/technology/vehicles.

Salinas, Pelayo. 2017. "Two Kick-Ass Speed Boats to Combat Illegal Shark Fishing in the Galápagos." GoFundMe, August 15, 2017. https://web.archive.org/web/20201021080021/https://charity.gofundme.com/o/en/team/two

-kick-ass-speed-boats-to-combat-illegal-shark-fishing-in-the-galapagos
/pelayosalinas.

Sandbrook, Chris. 2015. "The Social Implications of Using Drones for Biodiversity Conservation." *Ambio* 44 (4): 636–47.

Sandbrook, Chris, Rogelio Luque-Lora, and William M. Adams. 2018. "Human Bycatch: Conservation Surveillance and the Social Implications of Camera Traps." *Conservation and Society* 16 (4): 493–504.

Scarce, Rik. 1990. *Eco-warriors: Understanding the Radical Environmental Movement.* Chicago: Noble.

Schubel, Jerry, and Kimberly Thompson. 2019. "Farming the Sea: The Only Way to Meet Humanity's Future Food Needs." *GeoHealth* 3:238–44. https://doi
.org/10.1029/2019GH000204.

Schweitzer, Sarah. 2014. "Chasing Bayla." *Boston Globe*, October 26, 2014. https://
www3.bostonglobe.com/metro/2014/10/25/chasing-bayla/tJuazyjBOsd
KQTRVnAbh7K/story.html.

Scranton, Roy. 2015. *Learning to Die in the Anthropocene.* San Francisco: City Lights.

Sea Shepherd Conservation Society. 2012. "The M/V *Sam Simon* to Join the Sea Shepherd Fleet for the Next Voyage to Antarctica." Sea Shepherd Conservation Society website, June 21, 2012. https://web.archive.org/web
/20120624211308/http://www.seashepherd.org/news-and-media/2012/06/21
/the-mv-sam-simon-to-join-the-sea-shepherd-fleet-for-the-next-voyage-to
-antarctica-1395.

Sea Shepherd Conservation Society. 2017a. "Sea Shepherd Night Drone Shot Down by Poachers." Sea Shepherd Conservation Society website, December 26, 2017. https://web.archive.org/web/20210919090251/https://seashepherd
.org/news/sea-shepherd-night-drone-shot-down-by-poachers/.

Sea Shepherd Conservation Society. 2017b. "Sea Shepherd Night Drone Shot Down by Poachers." YouTube video, December 26, 2017. https://www.youtube
.com/watch?v=mmOZO7oI2Ws&ab_channel=SeaShepherdConservation
Society.

Sea Shepherd Conservation Society. 2018. "Operation Milagro IV: Sea Shepherd's Most Successful Vaquita Defense Campaign." YouTube video, June 11, 2018. https://www.youtube.com/watch?v=5AIRWMY-VKY&ab_channel=Sea
ShepherdConservationSociety.

Sea Shepherd Global. 2018. "Fleet Targeting Sharks in Timor Leste Released without Charge." Sea Shepherd Global website, July 4, 2018. https://www
.seashepherdglobal.org/latest-news/timor-leste-follow-up/.

Sea Shepherd UK. 2017. "Sea Shepherd Vessel Rammed by Fishing Boat in Panama." Sea Shepherd UK website, June 29, 2017. https://web.archive.org/web
/20200622100028/https://www.seashepherd.org.uk/news-and-commentary
/news/sea-shepherd-vessel-rammed-by-fishing-boat-in-panama.html.

Serafinelli, Elisa, and Lauren Alex O'Hagan. 2022. "Drone Views: A Multimodal

Ethnographic Perspective." *Visual Communication*. Published ahead of print, May 16, 2022. https://doi.org/10.1177/14703572211065093.

Serres, Michel. 2007. *Rameaux*. Paris: Le Pommier.

Shih, Douglas. 2015. "Orca Whale Pod San Juan Island Shih Photography." YouTube video, August 16, 2015. https://www.youtube.com/watch?v=_YjOyiBcfoc.

Simlai, Trishant. 2015. "Conservation 'Wars': Global Rise of Green Militarisation." *Economic and Political Weekly* 50 (50): 39–44. https://www.epw.in/journal /2015/50/perspectives/conservation-wars.html-0.

Sloterdijk, P. 2009. "Airquakes." *Environment and Planning D: Society and Space* 27 (1): 41–57.

Snitch, Thomas. 2015. "Satellites, Mathematics and Drones Take Down Poachers in Africa." *Conversation*, January 27, 2015. https://theconversation.com /satellites-mathematics-and-drones-take-down-poachers-in-africa-36638.

Society for Marine Mammalogy. N.d. "Guideline for the Treatment of Marine Mammals." Accessed May 5, 2023. https://marinemammalscience.org /about-us/ethics/marine-mammal-treatment-guidelines/.

Solly, Meilan. 2018. "Pacific Northwest Orca Population Hits 30-Year-Low." *Smithsonian*, July 10, 2018. https://www.smithsonianmag.com/smart-news /pacific-northwest-orca-population-hits-30-year-low-180969582/.

Spivak, Gayatri. 1988. "Can the Subaltern Speak?" In *Marxism and the Interpretation of Culture*, edited by Cary Nelson and Lawrence Grossberg, 271–313. Urbana: University of Illinois Press.

Squire, Rachel. 2016. "Rock, Water, Air and Fire: Foregrounding the Elements in the Gibraltar-Spain Dispute." *Environment and Planning D: Society and Space* 34, no. 3 (June): 545–63.

Srinivasan, Krithika. 2013. "The Biopolitics of Animal Being and Welfare: Dog Control and Care in the UK and India." *Transactions of the Institute of British Geographers* 38:106–19. https://doi.org/10.1111/j.1475-5661.2012.00501.x.

Srinivasan, Ramesh. 2019. *Beyond the Valley: How Innovators around the World Are Overcoming Inequality and Creating the Technologies of Tomorrow*. Cambridge, MA: MIT Press.

Srinivasan, Ramesh, and Adam Fish. 2009. "Internet Authorship: Social and Political Implications within Kyrgyzstan." *Journal of Computer-Mediated Communication* 14, no. 3 (April): 559–80. http://dx.doi.org/10.1111/j.1083-6101 .2009.01453.x.

Srinivasan, Ramesh, and Adam Fish. 2011. "Revolutionary Tactics, Media Ecologies, and Repressive Regimes." *Public Culture* 23, no. 3 (Fall): 505–10.

SR3. 2021. "SR3's Southern Resident Killer Whale Research Is Supporting Adaptive Conservation Measures." Sealife Response + Rehab + Research, September 17, 2021. https://www.sealifer3.org/news/sr3s-southern-resident-killer-whale-research-is-supporting-adaptive-conservation-measuresnbsp.

Starosielski, Nicole. 2015. *The Undersea Network*. Durham, NC: Duke University Press.

Starosielski, Nicole. 2021. "The Ends of Media Studies." *Public Culture* 33, no. 3 (September): 305–11.

Steffen, Will, Katherine Richardson, Johan Rockström, Sarah Cornell, Ingo Fetzer, Elena Bennett, Reinette Biggs, et al. 2015. "Planetary Boundaries: Guiding Human Development on a Changing Planet." *Science* 347, no. 6223 (February). http://doi.org/10.1126/science.1259855.

Steinberg, Philip E., Berit Kristoffersen, and Kristen L. Shake. 2020. "Edges and Flows: Exploring Legal Materialities and Biophysical Politics of Sea Ice." In *Blue Legalities: The Life and Laws of the Sea*, edited by Irus Braverman and Elizabeth R. Johnson, 85–106. Durham, NC: Duke University Press.

Steinberg, Philip, and Kimberley Peters. 2015. "Wet Ontologies, Fluid Spaces: Giving Depth to Volume through Oceanic Thinking." *Environment and Planning D: Society and Space* 33 (2): 247–64.

Sterne, Jonathan. 2022. *Diminished Faculties: A Political Phenomenology of Impairment*. Durham, NC: Duke University Press.

Stevenson, M. 2019. "Mexico: At Most Only 22 Vaquita Porpoises Remain." Associated Press, March 6, 2019. https://apnews.com/e7e8e2f5a3404644be020bf9deab60bc.

Stewart, Kathleen. 2011. "Atmospheric Attunements." *Environment and Planning D: Society and Space* 29 (3): 445–53. http://doi.org/10.1068/d9109.

Stiegler, Bernard. 1998. *Technics and Time: The Fault of Epimetheus*. Palo Alto, CA: Stanford University Press.

Stiegler, Bernard. 2018. *The Neganthropocene*. London: Open Humanities.

Sumaila, U. Rashid, Naazia Ebrahim, Anna Schuhbauer, Daniel Skerritt, Yang Li, Hong Sik Kim, Tabitha Grace Mallory, et al. 2019. "Updated Estimates and Analysis of Global Fisheries Subsidies." *Marine Policy* 109 (November): 103695.

Surf Life Saving Australia. 2021. *National Coastal Safety Report 2021*. September 7, 2021. https://issuu.com/surflifesavingaustralia/docs/ncsr_2021.

TallBear, Kim. 2011. "Why Interspecies Thinking Needs Indigenous Standpoints." *Fieldsights*, November 18, 2011. https://culanth.org/fieldsights/why-interspecies-thinking-needs-indigenous-standpoints.

Tate, R. D., B. P. Kelaher, C. P. Brand, B. R. Cullis, C. R. Gallen, S. D. A. Smith, and P. A. Butcher. 2021. "The Effectiveness of Shark-Management-Alert-in-Real-Time (SMART) Drumlines as a Tool for Catching White Sharks, *Carcharodon carcharias*, Off Coastal New South Wales, Australia." *Fisheries Management and Ecology* 28 (5): 496–506. https://doi.org/10.1111/fme.12489.

Taylor, Madeline, and Tina Soliman Hunter. 2020. "Equinor Has Abandoned Oil-Drilling Plans in the Great Australian Bight—So What's Next?" *Conversation*, February 26, 2020. https://theconversation.com/equinor-has-abandoned-oil-drilling-plans-in-the-great-australian-bight-so-whats-next-132435.

Thompson, Joanna. 2021. "A Drone Crash Caused Thousands of Elegant Terns to Abandon Their Nests." *Audubon*, June 11, 2021. https://www.audubon.org/news/a-drone-crash-caused-thousands-elegant-terns-abandon-their-nests.

Toonen, Hilde M., and Simon R. Bush. 2020. "The Digital Frontiers of Fisheries

Governance: Fish Attraction Devices, Drones and Satellites." *Journal of Environmental Policy and Planning* 22 (1): 125–37. https://doi.org/10.1080/15239
08X.2018.1461084.

Tschakert, Petra, David Schlosberg, Danielle Celermajer, Lauren Rickards, Christine Winter, Mathias Thaler, Makere Stewart-Harawira, et al. 2020. "Multispecies Justice: Climate-Just Futures with, for and beyond Humans." *WIRES Climate Change* 12, no. 2 (March/April): e699. https://doi.org/10.1002/wcc.699.

Tsing, Anna Lowenhaupt. 2015. *The Mushroom at the End of the World: On the Possibility of Life in Capitalist Ruins.* Princeton, NJ: Princeton University Press.

Tsing, Anna, Heather Swanson, Elaine Gan, and Nils Bubandt, eds. 2017. *Arts of Living on a Damaged Planet: Ghosts and Monsters of the Anthropocene.* Minneapolis: University of Minnesota Press.

Turner, Fred. 2006. *From Counterculture to Cyberculture: Stewart Brand, the Whole Earth Network, and the Rise of Digital Utopianism.* Chicago: University of Chicago Press.

Turnhout, Esther, Claire Waterton, Katja Neves, and Marleen Buizer. 2013. "Rethinking Biodiversity: From Goods and Services to 'Living With.'" *Conservation Letters* 6 (3): 154–61. https://doi.org/10.1111/j.1755-263X.2012.00307.x.

United Nations Environmental Program. 2018. "Intelligent Drones Crack Down on Illegal Fishing in African Waters." July 16, 2018. https://www.unep.org
/news-and-stories/story/intelligent-drones-crack-down-illegal-fishing
-african-waters.

Urbina, Ian. 2019. *The Outlaw Ocean: Journeys across the Last Untamed Frontier.* New York: Knopf.

Urbina, Ian. 2020. "The Deadly Secret of China's Invisible Armada." NBC News, July 22, 2020. https://www.nbcnews.com/specials/china-illegal-fishing-fleet/.

Valentine, Leonie E., Cristina E. Ramalho, Luis Mata, Michael D. Craig, Patricia L. Kennedy, and Richard J. Hobbs. 2020. "Novel Resources: Opportunities for and Risks to Species Conservation." *Frontiers in Ecology and the Environment* 18, no. 10 (September): 558–66. https://doi.org/10.1002/fee.2255.

Van der Leeuw, Sander. 2020. "The Role of Narratives in Human-Environmental Relations: An Essay on Elaborating Win-Win Solutions to Climate Change and Sustainability." *Climatic Change* 160 (4): 509–19.

Van Dooren, Thom. 2014. *Flight Ways: Life and Loss at the Edge of Extinction.* New York: Columbia University Press.

Van Dooren, Thom. 2019. *The Wake of the Crows: Living and Dying in Shared Worlds.* New York: Columbia University Press.

Van Dooren, Thom. 2020. "Story(telling)." *Swamphen* 7:1–2.

Van Dooren, Thom, Eben Kirksey, and Ursula Münster. 2016. "Multispecies Studies Cultivating Arts of Attentiveness." *Environmental Humanities* 8 (1): 1–23.

Van Dooren, Thom, and Deborah B. Rose. 2016. "Lively Ethnography: Storying Animist Worlds." *Environmental Humanities* 8 (1): 77–94.

Vaughan, Diane. 1997. *The Challenger Launch Decision.* Chicago: University of Chicago Press.

Vehlken, S., C. Vagt, and W. Kittler. 2021. "Introduction: Modeling the Pacific Ocean." *Media + Environment* 3 (2). https://doi.org/10.1525/001c.21997.

Vendl, C., E. Slavich, B. Wemheuer, T. Nelson, B. Ferrari, T. Thomas, and T. Rogers. 2020. "Respiratory Microbiota of Humpback Whales May Be Reduced in Diversity and Richness the Longer They Fast." *Scientific Reports* 10 (1): 1–13.

Verma, Audrey, René van der Wal, and Anke Fischer. 2016. "Imagining Wildlife: New Technologies and Animal Censuses, Maps and Museums." *Geoforum* 75:75–86. https://doi.org/10.1016/j.geoforum.2016.07.002.

Verran, Helen. 2009. "On Assemblage: Indigenous Knowledge and Digital Media (2003–2006), and HMS *Investigator* (1800–1805)." *Journal of Cultural Economy* 2 (1–2): 169–82.

Vertesi, Janet. 2012. "Seeing Like a Rover: Visualization and Embodiment on the Mars Exploration Rover Mission." *Social Studies of Science* 42 (3): 393–414.

Vincenot, Christian, and Sophie Petit. 2016. "Australia Too Casual with Protection Law." *Nature* 539 (November): https://doi.org/10.1038/539168d.

Virilio, Paul. 1999. *Politics of the Very Worst*. New York: Semiotext(e).

Wade, Lizzie. 2017. "Most Archaeologists Think the First Americans Arrived by Boat. Now, They're Beginning to Prove It." *Science*, August 10, 2017. https://www.science.org/content/article/most-archaeologists-think-first-americans-arrived-boat-now-they-re-beginning-prove-it.

Wade, Simeon. 2019. *Foucault in California [A True Story—Wherein the Great French Philosopher Drops Acid in the Valley of Death]*. Berkeley, CA: Heyday.

Wainwright, M. 2017. "Sensing the Airs: The Cultural Context for Breathing and Breathlessness in Uruguay." *Medical Anthropology* 36 (4): 332–47. https://doi.org/10.1080/01459740.2017.1287180.

Wang, Ning, Markus Christen, and Matthew Hunt. 2021. "Ethical Considerations Associated with 'Humanitarian Drones': A Scoping Literature Review." *Science and Engineering Ethics* 27 (4): 51. https://doi.org/10.1007/s11948-021-00327-4.

Waterton, Claire. 2010. "Barcoding Nature: Strategic Naturalization as Innovatory Practice in the Genomic Ordering of Things." *Sociological Review* 58:152–71. https://doi.org/10.1111/j.1467-954X.2010.01916.x.

Watson, Paul. 1981. *Sea Shepherd: My Fight for Whales and Seals*. New York: Norton.

Watson, Paul. 2015. "If the Ocean Dies, We All Die!" Sea Shepherd Conservation Society website, September 29, 2019. https://web.archive.org/web/20200918222205/https://seashepherd.org/2015/09/29/if-the-ocean-dies-we-all-die/.

Watson, Paul. 2019. "The Laws of Ecology, for the Survival of the Human Species." *LifeGate*, December 13, 2019. https://www.lifegate.com/paul-watson-laws-of-ecology.

Watson, Paul. 2020. *Earthforce! An Earth Warrior's Guide to Strategy*. 3rd ed. Self-published.

Watts, Vanessa. 2013. "Indigenous Place-Thought and Agency amongst Humans and Non-humans (First Woman and Sky Woman Go on a European World Tour!)." *Decolonization: Indigeneity, Education and Society* 2 (1): 20–34.

Weber, David. 2014. "WA Shark Cull: Activists Could Face Jail for Removing

Shark Bait." ABC News, January 27, 2014. https://www.abc.net.au/news/2014
-01-28/activists-could-face-jail-for-removing-shark-bait/5222052.

Weeks, Richard D., ed. 1992. *Global Shark Accident File.* http://sharkattackfile.net
/spreadsheets/pdf_directory/1922.02.04-Coughlan.pdf.

Werkheiser, Ian. 2015. "Fighting Nature: An Analysis and Critique of Breed-
Specific Flourishing Arguments for Dog Fights." *Society and Animals* 23, no. 5
(November): 502–20. https://doi.org/10.1163/15685306-12341375.

Werth, A. J., M. M. Kosma, E. M. Chenoweth, and J. M. Straley. 2019. "New
Views of Humpback Whale Flow Dynamics and Oral Morphology during Prey
Engulfment." *Marine Mammal Science* 35 (4): 1556–78. https://doi.org/10.1111
/mms.12614.

West, John G. 2011. "Changing Patterns of Shark Attacks in Australian Waters."
Marine and Freshwater Research 62, no. 6 (June): 744–54.

Weston, Kath. 2017. *Animate Planet: Making Visceral Sense of Living in a High-Tech
Ecologically Damaged World.* Durham, NC: Duke University Press.

Whale Wars. 2009a. "The Flexibility of Steel." Animal Planet, June 12, 2009.

Whale Wars. 2009b. "With a Hook." Animal Planet, July 17, 2009.

Whale Wars. 2010. "From Hell's Heart." Animal Planet, June 18, 2010.

Whitehead, Hal, Tim D. Smith, and Luke Rendell. 2021. "Adaptation of Sperm
Whales to Open-Boat Whalers: Rapid Social Learning on a Large Scale?" *Biological Letters* 17, no 3. https://doi.org/10.1098/rsbl.2021.0030.

Wigglesworth, Alex. 2021. "Drone Wipes Out Generation of Birds." *Los Angeles
Times*, June 8, 2021. https://enewspaper.latimes.com/infinity/article_share
.aspx?guid=18bf90ba-6581-4558-a73f-5b32557ccb9d.

Winner, Langdon. 1978. *Autonomous Technology: Technics-Out-of-Control as a Theme
in Political Thought.* Cambridge, MA: MIT Press.

Wolfe, Cary. 2012. *Before the Law: Humans and Other Animals in a Biopolitical
Frame.* Chicago: University of Chicago Press.

Worm, Boris, Brendal Davis, Lisa Kettemer, Christine Ward-Paige, Demian Chap-
man, Michael Heithaus, Steven Kessel, et al. 2013. "Global Catches, Exploitation Rates, and Rebuilding Options for Sharks." *Marine Policy* 40:194–204.

Wulf, Andrea. 2015. *The Invention of Nature: Alexander von Humboldt's New World.*
New York: Penguin Random House.

Youatt, Rafi. 2008. "Counting Species: Biopower and the Global Biodiversity Cen-
sus." *Environmental Values* 17 (3): 393–417.

Zak, Paul J. 2015. "Why Inspiring Stories Make Us React: The Neuroscience of
Narrative." *Cerebrum* 2 (February 2). https://www.dana.org/article/why
-inspiring-stories-make-us-react-the-neuroscience-of-narrative/.

Zimmer, Katarina. 2017. "'Ghost Poachers' Are Still at Large After the Big-
gest Shark-Smuggling Bust in Galápagos History." *Quartz*, August 24, 2017.
https://qz.com/1060639/galapagos-shark-fishing-bust-who-are-the-ghost-
poachers-who-supplied-the-fu-yuan-yu-leng-999/.

blue governmentality (*continued*)
elemental and legal limits, 27, 83; functionality, 26; legal contingency through
drone crashes, 25, 125–33, 135–36, 138–39;
limited functionality, 158; marine parks
and, 171–72; miscarriage of, 94–95; no-
take zones and high seas regulation, 171;
nonlethal shark mitigation, 154; ocean/
culture and, 132–33, 170, 171, 176, 184–
85, 187; paradoxes of, 16–17, 51, 71; role
of, 20–21; Sea Shepherd's attempts at,
54, 59, 61, 68, 71, 81, 87, 88; shark drone
program, 157, 160; storying and, 115–16;
as subjectivation, 89; subversion of, 95;
technophobic criticality and, 175
blue groper (*Acherodus*), 141
blue legalities, 16
blue whale (*Balaenoptera musculus*), 23, 29,
35, 72; exhale collection, 31; loudness of
breathing, 30, 42
Bolsa Chica Ecological Reserve, California, elegant tern eggs abandoned due to
crashing drone, 126–29, 135, 136, 138
Bondaroff, Teale, 24–25, 79–80
Bondi Beach, New South Wales: shark nets,
144; white sharks feeding near swimmers, 160
Borell, Andre, drone videography of
whale caught in shark net, Gold Coast,
151–54
bottlenose dolphin (*Tursiops aduncus*),
2, 10
Boucquey, Noëlle, 82, 159
Braidotti, Rosi, 4, 22
Bratton, Benjamin, 16–17, 45, 84, 89, 182,
183
Braverman, Irus, 15, 16, 173; criticism of
underwater drones, 174, 175; critique of
using underwater drones to kill crown-
of-thorns-starfish, 173
bridled tern (*Onychoprion anaethetus*), 165
"broken world thinking," 135
bronze whaler (*Carcharhinus brachyurus*),
141
brown bear (*Ursus arctos*), 161
Buck, Jim, 167–68, 170
bull shark (*Carcharhinus leucas*), 10, 145
Bunaken Island, 120
Bunaken Marine National Park in Northern Sulawesi, Indonesia, 119; drone mapping of coral bleaching, 119, 120
Büscher, Bram, 185
butyric acid bombs, 62, 68

"California Foucault," 13–14
Capricornia Cays National Park, Queensland, 164–72
carbon sequestration, 35
Cardno Report, 154, 155
care/control, 27, 47, 51, 90, 167–69, 171; blue
governmentality and, 87–89, 182
Castell, Manuel, 6
cetacean acoustic alarms, 157
cetacean science, 34–42
cetaceans: sensor attachment and tracking, 99; surveys, unpiloted aerial systems
use, 33
Challenger disaster, 133
China: extent of illegal, unreported, and
unregulated fishing, 73; handouts to fishing capitalists, 73; influence over smaller,
poorer, resource-rich neighbors, 78;
requested to check fishing fleet for violating fishing laws and CITES, 78–79
Chinese fishing fleet: shark finning:
Galápagos National Park, Ecuador, 74–75;
shark finning: Timor-Leste, 72–74, 75–79,
83, 88; whistleblower on *Fu Yuan Yu* vessel off Timor-Leste, 75–76, 88
Chinese river dolphin (*Lipotes vexillifer*), 35
Chinook salmon (*Oncorhynchus tshawytscha*), 130
Clark, Jonathan, 147–48, 149, 150, 162
climate change, 22, 88, 97, 104, 112, 120
climate science, 108, 182
coercive conservation, 81
coexistence: with sharks, 26, 151, 158–63;
with terrestrial predators, 161
"coexistence model," 161, 162
conservation, 84; colonialism behind, 85;
discontented humans and, 83–93; ethical
quandary of coral versus COTS for, 175;
limitations on, 88; needing support from
local communities, 85; ocean activists
and, 85–86; potential of blue governmentality, 187
conservation drones, 10, 84, 91, 139, 176;
of blue governmentality, 89, 91, 137, 188;
crashes. *See* crashing drones; for multispecies reworlding, 135–36; in multispecies studies, 136–38; piloting, 9; reverse
entropy and, 125. *See also* drone conservation; underwater drones
conservation geographers, 51, 83, 84, 88,
89–90
conservation technologies, 1–27, 48, 53, 54,
57, 84, 88; cultural geographers criticism

of, 89–90; drone use. *See* drone conservation; how they help nonhumans, 90; role in governing wild biopower, 185. *See also* specific types—e.g., helicopters

conservationists: drones and, 88; and local fisherfolk needs, 86; support for multispecies justice, 86–87

Convention on International Trade in Endangered Species of Wild Fauna and Flora (CITES), 79

convoluted neural networks (CNNs), 32–33

Coogee Beach, New South Wales, shark nets, 143, 144, 157

copper shark (bronze whaler) (*Carcharhinus brachyurus*), 141

coral, quandary in determining if coral or crown-of-thorns starfish more valuable for conservation, 175

coral bleaching, 4; Bunaken Marine National Park in Northern Sulawesi, Indonesia, 119, 120, 121; drone mapping of, 119, 120, 165, 170

coral/culture, 27, 164–72

coral destruction, 75

coral reefs, orthomosaic maps, 120, 165

Cousteau, Jacques, 106–7

COVID-19 global pandemic, 4, 16

Crandall, Jordan, 134

crash theory, 133–38; aerospace science and accidents, 133; cinema crashes, 133; conservation drone crashes as "events," 133–34; conservation drones reworlding through repair, 135–36; military drone crashes, 134–35; multispecies studies and conservation drones, 135, 136–38

crashing drones, 119–39, 179; avoiding crashes, 124; care and repair, 26; commonality of crashes, 122; coral reef, North Sulawesi, Indonesia, 120–21; elementality, 121–22; entropy and, 125, 129, 138–39; erupting Mount Agung volcano in Bali, 122; impact on nesting elegant terns, Southern California, 25, 126–29, 135, 136, 138, 177; impact on orca pod, Puget Sound, Washington State, 25, 126, 130–33, 135, 177; legal contingency through blue governmentality, 25, 125–32, 135–36, 138–39; legal reforms following, 134–35; metaphorical approaches, 124; pilots response to, 122, 124, 179; undersea fiber-optical cable, Iceland, 122, 123–24; videos of crashes into a range of objects, 127–28

crested tern (*Thalasseus bergii*) rookery, 168

crime and criminals, 50, 51, 59, 60, 78, 88, 95, 126, 188, 189. *See also* crashing drones; illegal fishing; poachers and poaching

Crist, Eileen, 84–85, 90

Crittercams, 112, 116; fixing to animals, 100–101; National Geographic footage from, 111; seal empowerment and, 110–11

crossbows and arrows, 36–37

crown-of-thorns starfish (COTS, *Acanthaster planci*): control, Lady Mulgrave Island, Queensland, 166, 168, 170, 173; current manual culling strategy, 173; RangerBots underwater drones to identify and kill, 173–74

cunjevoi (*Pyura stolonifera*), 141

Dalniy Vostok (Russian whaling ship), 57

data intimacy, 41–42, 112, 181

Dean, Jodi, 114

deep naturalism, 19, 59

deep-sea fishing by early humans, Timor-Leste, 143, 186

Deleuze, Gilles, 14, 159, 162

desert tortoise (*Gopherus agassizii*), protected from ravens by drones, Mojave Desert, 136–37

dingo (*Canis lupus dingo*), Indigenous Australians and, 137

diplomacy, biopolitics and, 136–37

direct actions: Sea Shepherd, 49–51, 53–54, 56, 58–59, 61, 62–64, 67–68; terrestrial, 64–65; to protect sharks, 146–47

direct enforcement, 79–81

Django, frees whale from shark net, Gold Coast, 152–53

Djibouti, US military drone crash, 134

dolphin, 2, 20, 35; radio tracking, 99

drone activism, 49–51, 59–62, 70, 151–53

drone cetology, 32–33, 40, 41; to collect exhale. *See* whale exhale

drone-collected images, 5

drone conservation, 1–27, 82, 84, 186–89; critiques, 12–13, 15; desert tortoise protection from ravens, Mojave Desert, 136–37; elementality, 11–12, 23, 24, 34, 45, 49–71, 178–80; intimacy, 6–7, 23–24, 25, 41–42, 48, 180–82; legally contingent efficacy, 94–95; potentialities and tests of, 94; Sea Shepherd use, 49–51, 59–62, 69, 70, 90; technicity, 7–10, 23–24, 28–48, 93, 94, 121, 176–78, 180; terrestrial use, 92; villagers' views on, 83–84 drone conservationists, 11–12, 15, 18, 92, 137

drone crashes. *See* crashing drones

drone data, 92, 104, 105; use as evidence of criminality in prosecutions, 79, 181, 187, 189

drone disconnection, 2

drone discourse, 90

drone disruption, avoiding, 136

drone failure. *See* crashing drones

drone heat-sensitive optics: detection of illegal totoaba fishing, 49–51; as tactical tool for Sea Shepherd, 50–51

drone images, machine learning informed by, 69

drone mapping: coral bleaching: Lady Musgrave Island, Queensland, 165; coral bleaching: Northern Sulawesi, Indonesia, 119, 120; undersea fiber-optical cables in Iceland, 122, 123–24

drone-microbe-whale intra-actions, 45–46

drone oceanographers, 10, 42, 116; blue governmentality and, 20

drone oceanography, 11, 23, 44, 59, 175, 184–85; comparison with crossbow for whale biopsies, 37; theorists' skepticism of, 102–3

drone piloting, 1–2, 3, 5; crashes and, 122, 124, 179; embodied acts of, 9; value judgments in, 93–94

drone regulations, 93–94, 130–31

Drone Shark App, 160

drone spectacle, 61

drone vision, 6, 46

drones: as affirmative multispecies ethics tool, 5; affordability, 71; automatic "return to home" function, 2, 26–27; bothering terrestrial animals, birds, and humans, 126–28, 136; crashing. *See* crashing drones; criticisms of, 92–93; difficult to recycle and unrepairable by design, 121–22; "direct enforcement," 79–80; as elevated network sensors, 120–21; engagement with the elements, 53; geofencing software, 93, 128; as intimate sensing systems, 7, 180; losing control of, 123–24; mediating data intimacy, 112; for nature surveillance, 26, 146, 154, 155, 160, 163, 188; orthomosaic mapping use, 120, 165, 170; posthuman vision, 39; positive and negative aspects, 86, 92–93; quantum physics and, 125; as remote-sensing

technologies, 180; technological achievements, 188; touching at a distance, 45; transforming oceans into a "frictionless field of data," 69. *See also* specific types— e.g., shark drones

drumlines (baited hooks), 144, 145–46, 154; activists and marine biologists experience of diving on, 149–50; activists checking, 147–49; freeing nonsharks found from, 149–50

Dunabin, Matt, 173

dusky flathead (*Platycephalus fuscus*), 140

dynamic ocean management, 159

eagles, incapacitation of drones, 127

Earth Live (documentary), 32, 33–34

earth, water, air, and fire, 10–11

Earth's ecosystem services, 87, 163

eastern reef egret (*Egretta sacra*), 169

Eco Shark Barrier, 154

ecocriticism, 85

ecological repair, 135

economic justice, 86

ecotourism, 89, 92

Ecuador, illegal shark fishing, 24, 74–75, 94

elegant tern (*Thalasseus elegans*): dislike of drones, 128; eggs abandoned due to drone's physical mimicry of predator, 126; impact of crashing drone on nesting colony, Southern California, 25, 126–29, 135, 136, 138, 177; legal protection of colonies from irresponsible drone operators, 128–29; loyalty to the flock, 126

elemental activism, 58, 60–61, 81

elemental media, 54–59

elemental thinking, 10–11

elementality, 10–12, 23, 24, 34, 45, 49–71, 178–79; atmoactivism, 62–65; atmospheric activism, 59–62; atmospheric and oceanic, 10, 11–12, 55, 83, 121, 151; crashed drones, 121–22; definition, 179–80; elemental media, 54–59; illegal totoaba fishing, 49–53; transelementality, 62, 65–68; transtechnologies, 62, 68–70. *See also under* drone conservation

elephant seal (*Mirounga*), 99

emperor penguin (*Aptenodytes forsteri*), 111

empowerment, ideology of, 110–11

entanglement, theories of, 150–51, 184

entropy, 135; crashing drones and, 125, 129, 138–39

environmental justice, 85, 125

humpback whale (*Megaptera novaeangliae*), 35, 111; association with other species, 47–48; blow circular "bubble nets," 44; caught in Queensland, 145; drone video showing whale caught and cut free from shark net, 151–53; exhale, 29; filming, 1–3, 4, 5; identification using AI, 32, 41; migration, eastern Australia, 5–6; song, 28, 34; viruses in whale-blow samples, 47, 48
Hunter, Robert, 57
Hutton, Jack, 49, 50–51
"hydro-logics," 55
hydrogen balloon cetology, 41

Iceland, undersea fiber-optical cables, 122, 123–24
Iggleden, Jason, 160
illegal fishing, 5, 24, 58, 59, 188; policing and punishment for, 188; shark finning, 24, 72–79, 83, 88, 177; totoaba, Sea of Cortez, Mexico, 49–51
Indian one-horned rhinoceros, Kaziranga National Park in Nepal, 92
Indigenous Gidarjil Bundaberg Land and Sea Rangers, monitoring Wallaginji (Lady Mulgrave Island) and Sea Country, 167
Intel, partnership with Ocean Alliance 32–33, 40
interior milieu, nature realism as, 52–54, 55, 56, 87
intimacy, 6–7, 23–24, 25, 41–42, 177, 180–82. *See also* data intimacy
intimate sensing systems, drones as, 7, 180

Jackson, Steven, 26, 135
Japanese Institute of Cetacean Research (ICR), 60, 62
Japanese whaling / whaling ships, 59–60, 62, 81, 85
Jiménez, Alberto Corsín, 143
Johnson, Elizabeth, 17

Kapp, Ernst, 38, 39
Keku Strait, Alaska, 32–34
Kerr, Iain, 40; collecting whale exhale, 29, 30, 31, 35; drone oceanography versus crossbow and arrows, 37
Kittler, Fredrich, 121
Knapp, Peter, 126
Koh, Lian Pin, 91–92

Kopnina, Helen, 24
Korotva, George, 57
Kuhn, Carey, 97–98, 99, 104, 111

laboratory analysis of drone-collected whale exhale, 46–48
Lady Elliot Island, Queensland, rewilding, 168
Lady Musgrave Island, Capricornia Cays National Park, 164–72; author's experience of navigating an underwater drone, 173–74; black noddies, bridled terns, and muttonbirds, rookery and noise, 165–66; blacktip reef sharks, 166; blue governmentality regulations on use, 170; campsite, 165, 166; care and control, 167–69, 171; crown-of-thorns starfish control, 166, 168, 170, 173; drone survey of coral bleaching / coral reefs, 165, 170; drone survey of turtle tracks, sea cucumber densities, and black-tip reef sharks, 170; Gidarjil Traditional Use of Marine Resources Agreement (TUMRA), 171; green sea turtles and hatchlings, 166, 167, 168–70; hawk moths, 166; Indigenous Gidarjil Bundaberg Land and Sea Rangers, 167; loggerhead sea turtle survey, 167; as protected island, 166–67; rewilding to remove exotic species, 168, 169; Traditional Owners' gathering and hunting, 166–67
laws: marine protection, 75–79, 129, 171, 187–88; wildlife protection, 130–31, 132, 134–35, 138–39
leaf blowers, as atmoactivist technology, 64–65
left governmentality, 14, 88–89
legal contingency within blue governmentality through drone crashes, 125–26; nesting elegant terns, Southern California, 25, 126–29, 136; orca pod, Puget Sound, Washington State, 126, 130–33, 138–39
legal reforms following drone crashes, 134–35
Lehman, Jessica, 69, 82, 102
leopard shark (*Triakis semifasciata*), 78
Leroi-Gourhan, André, 7, 23, 38–39, 71, 176, 178
Liboiron, Max, 8
little penguin (*Eudyptula minor*), 69
Loebl, Melissa, 126, 128
loggerhead sea turtle (*Caretta caretta*), 167

long-finned pilot whale (*Globicephala melas*), 59
Lorimer, Jamie, 16, 185
losing control of drones, 123–24
Lovelock, James, 139
Lovink, Geert, 113

McLuhan, Marshall, 8
macroplastic waste, 35
Malm, Andreas, 20
marine biodiversity decline, 139
marine biopower, 16, 188
marine life monitoring, drone use, 6
marine parks, 171–72
marine protection areas, 171; impact on local communities, 172
marine protection laws, 129, 171, 187–88
marine species decline, 4
Marshall, Greg, 100
mass extinction, sixth, 43, 87, 139, 184
Māui dolphin (*Cephalorhynchus hectori maui*), 35
media, 8, 26, 90, 91, 94, 107, 121, 153; elemental, 54–59. *See also* extinction media; sovereign media
media activists, 110
media biopolitics, 90–91
media studies scholars, 9, 10, 39, 44, 55, 61, 85, 107, 121, 134, 178, 188
mediation: as condition of embodiment, 45; elemental, 39, 45, 46; transelemental, 33, 34
Meinig, Chris, 104
methylcellulose, 62, 177
Mexican Armada (Navy), and Sea Shepherd, 50, 51, 59
Mexican surveillance drone crash in El Paso, Texas, 134
Mexicans, hostility toward the Armada and Sea Shepherd, 51
microbes, whales and, 32, 37, 42, 45, 46–48
microbiomes, whale, 47, 48
microplastic waste, 140, 187
military drone crash, 134–35
militarization of conservation, 4, 51, 88
Moffa, Anthony, 81
Mojave Desert, California, drones to protect desert tortoises from ravens, 135–37
Molsberry, Nick, 128
Mordy, Calvin, 104, 105, 114
Mount Agung volcano erupting in Bali, 122
mourning extinction, 4, 22, 137

Mullins, Russ, 130, 131, 132
multiple natures, 185
multispecies collaboration, storying as, 107–8, 109–110
multispecies ethnography, 138
multispecies justice, 86–87, 135
multispecies studies, 135, 136–38
Murphy, Michelle, 47
muttonbird (short-tailed shearwater) (*Ardenna tenuirostris*), 165, 166

Nahum-Claudel, Chloe, 144
National Geographic (TV channel), 32, 100, 111
Native American whaling, 137–38
natural technology, 39
nature constructivism, 21
nature realism, 17–19, 20, 52, 87, 185; direct action, 61; as "interior milieu," 52–54, 55, 56, 87
nature surveillance, drones for, 26, 146, 154, 155, 160, 163, 188
natureculturalism, 19, 21–23, 184–85, 189; critique of, 185; spectrum of, 21–22
natureculture, 18, 19, 185, 187
natureculture materialism, 21–22
negative biopolitics, 15, 84, 100, 156, 175, 188
Neimanis, Astrid, 55
nesting birds: crashing drone impact on elegant terns, 25, 126–29, 135, 136, 177; drone problem and, 128; drone use for counting and mapping, 128
net rippers, 58
neuston layer, 45
New South Wales (NSW): activists checking shark nets and drumlines, 147; protected shark species, 162
New South Wales Shark Management Strategy: bycatch, 145; net and drumline use on NSW beaches, 143, 144, 146, 147
Nisshin Maru (Japanese whale-meat processing ship), 60, 62
no-take (no-fishing) zones, 171
National Oceanic and Atmospheric Administration (NOAA): comparison of research vessel data collection efficiency with Saildrone, 101–2, 104; investigating northern fur seals diet, 25, 97; oceanographers preference of research vessel over drone, 103–4; pollock distribution influences seal diving behavior, 105; pollock fisheries and ecosystem status, 105–6;

National Oceanic and Atmospheric Administration (NOAA) (continued)
Saildrone data on seals not influencing fisheries management, 114–15; Saildrone and instrument tracking of northern fur seals, 15, 25, 96–103
North Atlantic right whale (Eubalaena glacialis), 4
northern fur seal (Callorhinus ursinus), 4, 25, 96–106; capture and Crittercam attachment and tracking, 98, 99, 100–101; energy effort in pursuit of prey, 105; impact of warming sea and Arctic ice recession on, 104, 105; mother seal capture, 97–98; recapture and Crittercam removal, 105; St. Paul Island, Alaska, 96, 100–101; subsurface world from a fur seal's point of view, 105; tracking using Saildrones, 96–97, 100–103; vulnerability, 97; walleye pollock in diet of, 25, 97, 105
notworking of networked communication, 110, 112–15
noumena, 21, 22

ocean acidification, 120, 188
ocean activists and activism, 5, 57–59, 65–67, 85–87, 177, 183, 184; checking shark nets and drumlines, 147–49; confronting shark mortality, 150–51; experience of diving on drumlines, 149–50. See also specific organizations—e.g. Sea Shepherd Conservation Society
Ocean Alliance, 23, 28, 44, 71; collecting whale exhale/microbiota, 29–32, 41, 46, 48; drone cetology, 41; hearing whales, 28–29; individualizing whales through photography, Patagonia, 40–41; Intel AI whale identification system, 32–33, 40
ocean conservation, 15
ocean/cultures, 19–23, 27, 47, 111, 150–51, 158, 167, 171; blue governmentality and, 132–33, 170, 171, 176, 184–85, 187; shark drones monitoring boundaries of, 161–62. See also oceancultures
ocean drones, 93, 112, 176; NOAA oceanographers views on, 103–4; theorists' skepticism of, 102–3. See also Saildrones; underwater drones
ocean materiality, 55, 66, 68, 82
Ocean Warrior (Sea Shepherd vessel), 75, 76
oceancultures, 111, 158, 171, 187

oceanic biomass, factors influencing, 172, 188
oceanic elementality 10, 11–12, 83, 121, 151
oceanic geontopower, 82
oceaning practice, 20–21
oceanographic fieldwork, 69
oceans: economic dependence on, 186; human interactions with, 186–87; "more-than-wet ontology," 68; protein sources, 188
Oceans Asia, 79
O'Grady, Nathaniel, 82
oil exploration, Falkland Islands, 100
open-water swimmers, sharks and, 161
Operation Jeedara (Sea Shepherd), 69
Operation Milagro (Sea Shepherd), 49–51, 59
Operation Zero Tolerance (Sea Shepherd), 59–60
Oppermann, Serpil, 25, 109–110
orca (Orcinus orca): animal tagging, 99; drone for pregnancy monitoring, 130, 131; illegal close deployment of drones for filming, Puget Sound, 130, 131–32; photographic population monitoring, Puget Sound, 39–40; Puget Sound, 4, 39, 45, 130; stopping drone crashes on orca pods, Puget Sound, Washington State, 25, 126, 130–33, 135, 139–40, 177; Washington State proposed legislation over drone use, 130–31, 132; wildlife health monitoring technologies, wildlife protection, and drone phenomena, 132
originary technicity, 9, 38, 43, 46, 88, 1881
orthomosaic maps: coral reefs, 120, 165, 170; turtle nesting pathways and sea cucumber densities, 170
overfishing, 73, 87, 97, 112, 115, 130

Pacific cod (Gadus macrocephalus), 104
Pacific salmon (Oncorhynchus), 104
parachute cetology, 40, 41
Patagonia, Chile, 40–41
Payne, Roger: Songs of the Humpback Whale (album), 28, 34, 57; whale individualizing, Patagonia, Chile, 40–41
penguin, 69, 100, 111
peregrine falcon (Falco peregrinus), 126
Peters, John Durham, 44–45, 54–55
Peters, Kimberley, 24, 57
petri dishes: collecting whale snot on, 29, 31; delays analyzing snot samples, 46;

viruses in whale-blow samples, 47; whale microbes transferred to, 45–46

phenomenological approach, 9

phytoplankton, 55, 101, 186

pisonia forests (*Pisonia grandis*), Lady Musgrave Island, Queensland, 164, 168

plastic waste, 35, 140

poachers and poaching, 9, 20, 24, 59, 84, 85, 91, 189; decriminalizing, 84; drone documentation, 60; effect on local people, 84; shark finning: Galápagos National Park, Ecuador, 74–75; shark finning; Timor-Leste, 72–74, 75–79, 83, 88; strategies to limit, 92; terrestrial use, 91–92; totoaba fishing, 49–51; use of automatic weapons to down drones, 51, 91

Policia Nacionale Timor-Leste (PNTL), works with Sea Shepherd over illegal fishing, 73, 75–78, 88

political decision-making, storying to inform, 108–9

political natureculturalism 21, 22

political solutionism, 84–85

politics of subjectivity, 90–91

pollock. *See* walleye pollock (*Gadus chalcogrammus*)

pollutants/pollution, 32, 42, 46, 87, 130, 140, 188

population health studies, 69

porpoise killing, 24, 49, 57

positive biogovernmentality, 16–17

positive biopolitics, 17, 100, 175, 182

posthuman vision, drone's, 39

Povinelli, Elizabeth, 81–82

primeval marine traps, 143–44

prop-foulers, 66–67, 177

prosecutions, drone date for, 79, 181, 187, 189

protest activism, 80, 81

Puget Sound orca, 4, 39–40, 45; drone monitoring to predict pregnancy, 130; factors in decline, 130; protecting from drone crashes, 126, 130–33, 135, 139–40

quantum physics, drones and, 125

Queensland: activists monitoring shark nets and drumlines, 147–48; drone surveillance of sharks, southern Queensland beaches, 154; efficacy of shark net control, 162; nonlethal shark barriers and deterrents, north Queensland beaches, 154; philosophy of coexistence with sharks, 154–55; SharkSmart Drone Trial, 155; videography of humpback whale caught in shark net, 151–54

Queensland Marine Animal Release Team, 149, 152

Queensland Shark Control Program: number and species of sharks caught, 145; shark nets and baited drumline use, 144, 145–46

radical environmentalists, 54

radio tracking of cetaceans, 99

RangerBots underwater drones, 174, 175; use of AI to identify and kill crown-of-thorns starfish, 173

raven (*Corvus corax*), frightened by drones away from desert tortoises, Mojave Desert, 136

reciprocities, stories as, 108

red fox (*Vulpes vulpes*) control, 167

Redrobe, Karen, 133, 135

relative technicity, 8–9, 23, 27, 54, 94

remora (*Remora*), 100

remote-sensing systems, 68–69, 86, 180

respiration, whales', 32, 42–46

response-ability, 112

rewilding, Queensland islands, 168, 169

reworlding through repair, 125, 135–36

Richardson, Joanna, 113

Richmond, Holly, 149–50

rigid inflatable boats (RIBs), 57, 60, 66–68

Rogan, Andy, 28, 29, 32

rorqual whale (*Balaenopteridae*), 101

Rose, Deborah Bird, 25, 107, 137

Russian whalers, 56, 57

Saildrones, 96–97, 101, 112, 116–17, 121; benefits of, 103; comparison with massive NOAA vessels for data collection, 101–2, 103–4; data not integrated into fisheries management, 114; echolocation of walleye pollock, 97; investigating the edge of Arctic Sea ice, 104; in seal research, as "basic research," 113; survey grid, 103; theorists' skepticism of, 102–3; tracking of northern fur seals through Bering Sea, 97, 100–103, 105

St. Paul Island, Alaska, mother northern fur seals fitted with Crittercams, 96, 100–101

Salinas, Pelayo, 74

Salish Sea orca, 39

Sandbrook, Chris, 12–13, 84–85
sawfish (*Pristis* spp.), 24
scalloped hammerhead (*Sphyrna lewini*), 24
science as activism, 46
scientific story work, 108
Sea of Cortez, Mexico, 24; collecting whale exhale, 28–31; illegal totoaba fishing, 49–51
Sea Country, aerial and underwater drone monitoring, 167
sea cucumber, 170
"sea epistemology," 55
Sea of Shadows (documentary), 51, 52, 59
Sea Shepherd Conservation Society, 24; activism benefits, 87; air and ocean materiality, 55, 66, 68; assists Mexican Armada (Navy) to interdict poaching, 50–51, 59; atmoactivism, 62–64, 70; atmospheric activism, 59–62; blue governmentality, 54, 59, 61, 68, 71, 81, 87, 88; collaborative approach with institutions, 69; criticized for ignoring the rights of local fisherfolk, 86; direct actions, 49–51, 53–54, 56, 58, 61, 62–65, 67–68; direct enforcement, 79–81; drone conservation use, 50–51, 52, 59–62, 69, 70, 75–76, 90, 91; drone shot out of the sky, 51, 91; as existential technology for endangered species and volunteers, 57–58; harp seal hunting, ice floes, Labrador, 63–64; helicopter use, 52, 60, 62–64, 67, 68; media biopolitics, 90–91; nature realism as an "interior milieu," 52–54, 55, 56, 87; on-board sound and communication technologies, 56–57; Operation Jeedara (counting Australian sea lions and little penguins), 69; Operation Milagro campaign (against illegal totoaba fishing), 49–51, 59; Operation Zero Tolerance (against Japanese whaling), 59–60, 81; purpose, 90–91; pursuit of Russian and Japanese whalers, 56–57, 59–69; shark finning: Ecuador, 74–75; shark finning: Timor-Leste, 72–74, 75–79, 88; ships rammed, 80, 91; ships and RIBs, 52, 56, 57–58, 66–68, 90, 91; transelementalities, 66–68; transtechnologies, 69; volunteers, 57–58, 70, 85; witnessing impressive spectacles, 61; works with PNTL over illegal fishing, 73, 75–78
sea slug (*Nudibranchia*), 141
sea surface microlayer, 45
seal hunting, Labrador, 63–64, 99

seal stories, 25, 107–8, 109–110, 113–14, 115; telling, 116–18
seals: capture and sensor fitting, 98–99, 112; empowerment via Crittercams, 110–11; in extinction media, 113–14; future Saildrome interaction with, 117; not benefiting from research, 113–14; Saildrome data and fisheries management, 105, 114–15, 117; scientists role in storying about, 115, 116–18; unintentional participation in oceanographic sampling, 109–110, 114. *See also* northern fur seal (*Callorhinus ursinus*)
seascape epistemology, 11, 39
sense-ability, 112
"separation model," 161
Serres, Michel, 39
shark barriers: nonlethal, 154, 155. *See also* shark nets
shark bites, 26, 142; Australia, 142–43, 145–46; worldwide increase, 142
shark control: drumline (baited hooks) use, 26, 144, 145–46, 147–50, 154; efficacy, 145–46, 162. *See also* shark barriers; shark nets
shark drones: applications, 156, 179; automation, 156; benefits for shark mitigation, 155–56; blue governmentality and, 157, 160; community support for, 155, 156; fluid lensing software and AI to boost "seeing" below the water's surface, 155–56; monitoring boundaries of ocean/cultures, 161–62; not suitable for northern Queensland beaches due to water opacity, 154; recommended for southern Queensland beaches, 154; SharkSmart Drone Trial, 155; to allow coexistence, 159–60, 162–63; trialed to identify sharks and alert swimmers, 146, 154, 155, 160, 163
shark finning, 177; Galápagos National Park, Ecuador, 74–75; Timor-Leste, 72–74, 75–79, 83, 88
shark/human coexistence, 26, 151, 158–63
shark/human connection, 148–51, 159
shark nets, 26, 143, 146, 151, 177, 179; activists checking, 147–49; countries using, 144; drone video of humpback whale caught and cut free, Gold Coast, 151–53; efficacy, 162; New South Wales beaches, 143, 144, 147, 157; objective of, 144–45; operation, 145; Queensland beaches, 144, 145–46, 147–48, 157; removal during

United Nations Convention on the Law of the Sea, 171
unpiloted aerial systems (UAS), 33
urchin (*Echinoidea*), 141
US National Oceanic and Atmospheric Administration. *See* National Oceanic and Atmospheric Administration (NOAA)

Van Dooren, Thom, 25, 107, 136–37
vaquita porpoise (*Phocoena sinus*), 35, 49–50, 51
Verran, Helen, 46
viral clouds, whale, 47–48

walleye pollock (*Gadus chalcogrammus*), 25, 97; American fisheries policy, 105–6, 117; Bering Sea population, 97; highly desired by humans, 97; impact of Bering Sea warming on, 104, 115; NOAA annual stock assessment, 105; in northern fur seal diet, 97, 105; overfishing, 97, 115
Washington State: orcas and drone tourism, 131; proposal to ban flying drones over orcas, Puget Sound, 130–31
water cannons, 62, 67, 77, 177, 184
water pollution, Sydney, New South Wales, 140
Watson, Paul, 25, 52, 54, 55, 61, 62, 80, 86
Watts, Vanessa, 138
wedge-tailed eagle (*Aquila audax*), attack flying drones, 127
Western Australia: beachgoer encounters with sharks, 161; No WA Shark Cull campaign, 146–47
Weston, Kath, 6, 107
wet wilderness, ocean as, 94–95
whale biopsies, 36–37
whale breathing/respiration, 32, 42–46
whale bubbles, 44
whale ecotourism, 89
whale exhale, 1, 23, 24, 28–29, 37, 42, 177; analyzing to understand health, 43, 46, 47; bacteria, viruses, and hormones in, 46–47; drone collecting, 29–31, 34, 42; species identification by shape of exhale, 29. *See also* whale snot

whale fluke identification, 2, 32–33
whale health, 23, 43, 47
whale identification, 6; from helicopters, planes, parachutes, and balloons, 40–41, 177; from shape of exhale, 29; using real-time artificial intelligence (AI), 32–33, 40, 41, 177
whale lungs, 42–43
whale microbes, 32, 37, 42, 45, 46–48
whale photography, 39–41
whale poaching, 24
whale shark (*Rhincodon typus*), 59
whale snot, 29–31, 34, 46–47, 48
whale songs, 57
whale watching, 89
whaler watching organizations, 64
whaler sharks (*Carcharhinidae*), 146
whales: carbon sequestration by rehabilitating global whale populations to prewhaling levels 35–36; drone use for monitoring and studies, 1–3, 32–34, 40–41, 177; "ecological service," 35–36; exposure to human-generated chemicals, 46–47; field research with, 41–42; living in the Blue Anthropocene, 35; microbiomes, 47, 48; neuston layer and, 45; portrayal in environmental activism, 34–35; sampling and tagging, 36–37; as "sentinel species" of ocean sustainability, 35–36; shark nets and, 151–53
whaling, 56, 57, 58, 59, 62–63, 81; Native American, 137–38. *See also* antiwhaling activists
whaling moratorium, 4, 35, 60
whaling ships, 56, 57, 66, 177
white shark (*Carcharodon carcharias*), 10, 145, 146, 160
white-striped octopus (*Callistoctopus ornatus*), 169
wildlife protection, 92
wildlife protection laws, 130–31, 132, 138–39
Wolfe, Cary, 183
wolverine (*Gulo gulo*), 161

yellowtail kingfish (*Seriola lalandi*), 2